Giovanni Gallavotti (Ed.)

Statistical Mechanics

Lectures given at a Summer School of the
Centro Internazionale Matematico Estivo (C.I.M.E.),
held in Bressanone (Bolzano), Italy,
June 21-27, 1976

Springer

FONDAZIONE
CIME
ROBERTO CONTI

C.I.M.E. Foundation
c/o Dipartimento di Matematica "U. Dini"
Viale margagni n. 67/a
50134 Firenze
Italy
cime@math.unifi.it

ISBN 978-3-642-11107-5 e-ISBN: 978-3-642-11108-2
DOI:10.1007/978-3-642-11108-2
Springer Heidelberg Dordrecht London New York

Reprint of the 1st ed. C.I.M.E., Ed. Liguori, Napoli 1976
With kind permission of C.I.M.E.

Printed on acid-free paper

Springer.com

CENTRO INTERNAZIONALE MATEMATICO ESTIVO

(C.I.M.E.)

I Ciclo - Bressanone dal 21 giugno al 24 giugno 1976

STATISTICAL MECHANCIS

Coordinatore: Prof. Giovanni Gallavotti

P. CARTIER

Thèorie de la misure

Introduction à la mécanique statistique classique.

(Testo non pervenuto)

CENTRO INTERNAZIONALE MATEMATICO ESTIVO

(C.I.M.E.)

A SKETCH OF THE THEORY OF THE BOLTZMANN EQUATION

C. CERCIGNANI

Istituto di Matematica, Politecnico di Milano

Corso tenuto a Bressanone dal 21 giugno al 24 giugno 1976

A Sketch of the Theory of the Boltzmann equation

Carlo Cercignani
Istituto di Matematica
Politecnico di Milano
Milano, Italy

In this seminar, I shall briefly review the theory of the Boltzmann equation. How the latter arises from the Liouville equation has been discussed in O.Lanford's lectures.

We shall write the Boltzmann equation in this form

$$\frac{\partial f}{\partial t} + \underline{\xi} \cdot \frac{\partial f}{\partial \underline{x}} = Q(f,f) \tag{1}$$

where t, $\underline{x}$, $\underline{\xi}$ denote the time, space and velocity variables, while f is the distribution function, normalized in such a way that

$$\int f \, d\underline{x} \, d\underline{\xi} = M \tag{2}$$

where M is the mass contained in the region over which the integration with respect to $\underline{x}$ extends.

$Q(f,f)$ is the so called collision term, explicitly obtainable from the following definition

$$Q(f,g) = \frac{1}{2m} \int (f' g'_* + f'_* g' - f g_* - f_* g) B(\theta, V) d\underline{\xi}_* d\theta d\varepsilon \tag{3}$$

where $\underline{\xi}_*$ is an ausiliary velocity vector, V is the relative speed, i.e. the magnitude of the vector $\underline{V} = \underline{\xi} - \underline{\xi}_*$, $f' = f(\underline{\xi}')$, $g'_* = g(\underline{\xi}'_*)$ etc., where $\underline{\xi}'$ and $\underline{\xi}'_*$ are related to $\underline{\xi}$ and $\underline{\xi}_*$ through the relations expressing conservation of momentum and energy in a collision

$$\underline{\xi}' + \underline{\xi}'_* = \underline{\xi} + \underline{\xi}' \qquad (4)$$

$$\xi'^2 + \xi'^2_* = \xi^2 + \xi'^2 \qquad (5)$$

equivalent to

$$\underline{\xi}' = \underline{\xi} - \underline{n}\,(\underline{n} \cdot \underline{V}) \qquad (6)$$

$$\underline{\xi}'_* = \underline{\xi}_* + \underline{n}\,(\underline{n} \cdot \underline{V}) \qquad (7)$$

where $\underline{n}$ is a unit vector, whose polar angles are θ and ε in a polar coordinate system with $\underline{V}$ as polar axis. Integration extends to all values of $\underline{\xi}_*$ and between 0 and $\pi/2$ with respect to θ, from 0 to 2π with respect to ε. Finally $B(\theta, V)$ is related to the differential cross section $\sigma(\theta, V)$ by the relation

$$B(\theta, V) = V \sin\theta\, \sigma(\theta, V) \qquad (8)$$

and m is the mass of a gas molecule. For further details one should consult one of my books [1,2].

Eq. (1) is valid for monatomic molecules and is more general than the Boltzmann equation considered by Lanford in his lectures, because it is not restricted to rigid spheres, but allows molecules with any differential cross section. The case of rigid spheres is obtained by specializing $B(\theta, V)$ as follows

$$B(\theta,V) = V d^2 \sin\theta \cos\theta \qquad (9)$$

where d is the sphere diameter. Another important case is offered by the so called Maxwell molecules.

The latter are classical point masses interacting with a central force inversely proportional to the fifth power of their mutual distance; as a consequence, it turns out that $B(\theta,V)$ is independent of V.

It is clear that initial and boundary conditions are required in order to solve the Boltzmann equation, since the latter contains the time and space derivatives of f. The boundary conditions are particularly important since they describe the interaction of the gas molecules with solid walls, but particular difficult to establish; the difficulties are due, mainly, to our lack of knowledge of the structure of the surface layers of solid bodies and hence of the interaction potential of the gas molecules with molecules of the solid. When a molecule impinges upon a surface, it is adsorbed and may form chemical bonds, dissociate, become ionized or displace surface atoms.

The simplest possible model of the gas-surface interaction is to assume that the molecules are specularly reflected at the solid boundary. This assumption is extremely unrealistic in general and can be used only in particular cases. In general, a molecule striking a surface at a velocity $\underline{\xi}'$ reflects from it at a velocity $\underline{\xi}$ which is strictly determined only if the path of the molecule within a wall can be computed exactly. This computation is impossible because it depends upon a great number of details, such as the locations and velocities of all the molecules of the wall. Hence we may only hope to compute the probability density $R(\underline{\xi}' \longrightarrow \underline{\xi})$ that a molecule striking the surface with velocity between $\underline{\xi}'$ and $\underline{\xi}'+d\underline{\xi}'$ re-emerges with velocity between $\underline{\xi}$ and $\underline{\xi}+d\underline{\xi}$. If R is known, it is easy to write the boundary condition for f [1,2] :

$$|\underline{\xi}\cdot\underline{n}|\, f(\underline{x},\underline{\xi},t) = \int_{\underline{\xi}'\cdot\underline{n}<0} R(\underline{\xi}'\rightarrow\underline{\xi})\, f(\underline{\xi}')\, |\underline{\xi}'\cdot\underline{n}|\, d\underline{\xi}' \tag{10}$$

where $\underline{n}$ is the unit vector normal to the wall and we assumed the wall to be at rest (otherwise $\underline{\xi}$, $\underline{\xi}'$ must be replaced by $\underline{\xi}-\underline{u}_o$, $\underline{\xi}'-\underline{u}_o$, $\underline{u}_o$ denoting the wall's velocity.)

In general, R will be different at different points of the wall and different times; the dependence on $\underline{x}$ and t is not shown explicitly to make the equations shorter.

If the wall restitutes all the gas molecules (i.e. it is non-porous and nonadsorbing), the total probability for an impinging molecule to be re-emitted, with no matter what velocity $\underline{\xi}$, is unity:

$$\int_{\underline{\xi}\cdot\underline{n}} R(\underline{\xi}'\rightarrow\underline{\xi})\, d\underline{\xi} = 1 \tag{11}$$

An obvious property of the kernel $R(\underline{\xi}'\rightarrow\underline{\xi})$ is that it cannot assume negative values

$$R(\underline{\xi}'\rightarrow\underline{\xi}) \geq 0 \tag{12}$$

Another basic property of the kernel R , which can be called the "reciprocity law" or the "detailed balance", is written as follows [1, 2] :

$$|\underline{\xi}'\cdot\underline{n}|\, f_o(\underline{\xi}')\, R(\underline{\xi}'\rightarrow\underline{\xi}) = |\underline{\xi}\cdot\underline{n}|\, R(-\underline{\xi}\rightarrow-\underline{\xi}')\, f_o(\underline{\xi}) \tag{13}$$

where $f_o(\underline{\xi})$ is proportional to $\exp[-\xi^2/(2RT_o)]$, where T_o is the temperature of the wall (in other words, $f_0(\underline{\xi})$ is a Maxwellian distribution for a gas at rest at the temperature of the wall).

We note a simple consequence of reciprocity; if the impinging distribution is the wall Maxwellian f_0 and mass is conserved at the wall according to Eq.(11), then the distribution function of the emerging molecules is again f_0 or, in other words, the wall Maxwellian satisfies the boundary conditions. In fact, if we integrate Eq. (13) with respect to $\underline{\xi}'$ and use Eq. (11) we obtain

$$\int_{\underline{\xi}'\cdot\underline{n}<0} |\underline{\xi}'\cdot\underline{n}|\, f_0(\underline{\xi}')\, R(\underline{\xi}'\to\underline{\xi})\, d\underline{\xi}' = |\underline{\xi}\cdot\underline{n}|\, f_0(\underline{\xi}) \qquad (\underline{\xi}\cdot\underline{n}>0) \tag{14}$$

and this equation proves our statement, according to Eq. (10). It is to be remarked that Eq. (14), although a consequence of Eq. (13) (when Eq. (11) holds) is less restrictive than Eq. (13) and could be satisfied even if Eq. (13) failed.

As a consequence of the above properties, one can prove [2] the following remarkable theorem:

Let $C(g)$ be a strictly convex continuous function of its argument g. Then for any scattering kernel $R(\underline{\xi}'\to\underline{\xi})$ satisfying Eqs. (11), (12), (14), the following inequality holds

$$\int f_0\, \underline{\xi}\cdot\underline{n}\, C(g)\, d\underline{\xi} \le 0 \tag{15}$$

where f_0 is the wall Maxwellian, $g = f/f_0$ and integration extends to the full ranges of values of the components of $\underline{\xi}$, the values of f for $\underline{\xi}\cdot\underline{n}>0$ being related to those for $\underline{\xi}\cdot\underline{n}<0$ through Eq. (1.6). Equality in Eq. (15) holds if and only if $f=g$ almost everywhere, unless $R(\underline{\xi}'\to\underline{\xi})$ is proportional to a delta function.

As a corollary, the following inequality holds [2] :

$$\int \underline{\xi}\cdot\underline{n}\, f \log f \, d\underline{\xi} \leq -\frac{1}{RT_0}[\underline{q}\cdot\underline{n}]_{solid} \qquad (16)$$

where $[\underline{q}\cdot\underline{n}]_{solid}$ denotes the normal heat flux fed into the gas by the solid constituting the wall and R is the gas constant.

We want to generalize the H-theorem, considered in O. Lanford's lectures, to the case of a gas bounded by solid walls which may or may not be at rest. To this end we define

$$\mathcal{H} = \int f \log f \, d\underline{\xi} \qquad (17)$$

$$\mathcal{H}_i = \int \xi_i f \log f \, d\underline{\xi} \qquad (i=1,2,3) \qquad (18)$$

and observe that

$$\frac{\partial \mathcal{H}}{\partial t} + \frac{\partial \mathcal{H}_i}{\partial x_i} = \int \log f \; Q(f,f)\, d\underline{\xi} \qquad (19)$$

(A sum with respect to i from 1 to 3 is understood).

Now, the following identity holds for any φ, f, g provided the integrals make sense:

$$\int \varphi \, Q(f,g)\, d\underline{\xi} = \frac{1}{8m}\int (\varphi + \varphi_* - \varphi' - \varphi'_*)(f'g'_* + f'_*g' - fg_* - f_*g)\, B(\theta,V)\, d\underline{\xi}_*\, d\theta\, d\varepsilon \qquad (20)$$

This identity follow by straightforward manipulations; for details, see [1,2] .

Applying Eq. (20) to the case $\varphi = \log f,\ g = f\ (f \geq 0)$, we obtain

$$\int \log f \; Q(f,f)\, d\underline{\xi} = \frac{1}{8m}\int \left(\log \frac{ff_*}{f'f'_*}\right)\left(1 - \frac{ff_*}{f'f'_*}\right) f'f'_*\, B(\theta,V)\, d\underline{\xi}_*\, d\theta\, d\varepsilon \leq 0 \qquad (21)$$

where the inequality follows from the fact that $(1-\lambda)\log\lambda$ is always negative, except for $\lambda = 1$, where it is zero. Hence equality in Eq. (21) is valid if and only if

$$ff_* = f'f'_* \tag{22}$$

or letting φ denote $\log f$

$$\varphi + \varphi_* = \varphi' + \varphi'_* \tag{23}$$

This equation is satisfied trivially by $\varphi = 1$ and, as a consequence of Eqs. (4) and (5), by $\varphi = \xi_i$ (i = 1,2,3) and $\varphi = \xi^2$; it can be shown [2] that there a no other linearly independent collision invariants (such is the name for the solutions of Eq. (23)). As a consequence, the most general distribution function satisfying Eq. (22) is given by

$$f = \exp(a + \underline{b}\cdot\underline{\xi} + c\xi^2) \tag{24}$$

where a, $\underline{b}$, c are constant. Eq. (24) can be rewritten in the following form

$$f = \rho(2\pi RT)^{-3/2}\exp\left[-\frac{(\underline{\xi}-\underline{v})^2}{2RT}\right] \tag{25}$$

where ρ, $\underline{v}$, T are new constants related to the previous ones and have the meaning of density, mass velocity and temperature associated with the distribution function f according to well-known formulas [1,2]. Eq. (25) gives a Maxwellian distribution.

Eqs. (19) and (21) imply that

$$\frac{\partial \mathcal{H}}{\partial t} + \frac{\partial \mathcal{H}_i}{\partial x_i} \le 0 \tag{26}$$

where the equality sign applies if and only if f is Maxwellian, i.e. is given by Eq. (25).

If we integrate, both sides of Eq. (26) with respect to $\underline{x}$ over a region R bounded by solid walls, we have, if the boundary ∂R of R moves with velocity $\underline{u}_0$:

$$\frac{dH}{dt} - \int_{\partial R} (\underline{\mathcal{H}} \cdot \underline{n} - \mathcal{H} \underline{u}_0 \cdot \underline{n}) \, dS \leq 0 \tag{27}$$

where dS is a surface element of the boundary ∂R and $\underline{n}$ the inward normal. The second term in the integral comes from the fact that, if the boundary is moving, when forming the time derivative of H we have to take into account that the region of integration changes with time.

If we use Eq. (16), Eq. (27) becomes:

$$\frac{dH}{dt} \leq -\frac{1}{R} \int \frac{(q \cdot \underline{n})_{solid} \, dS}{T_0} \tag{28}$$

where we replaced $\underline{\xi}$ by $\underline{\xi} - \underline{u}_0$ in Eq. (16) as required. Eq. (28) generalizes the H-theorem, showing that H decreases with time if there is no heat exchange between the gas and the walls. Also, equality in Eq. (28) applies if and only if f is Maxwellian. Eq. (28) suggests that H be interpreted as $-\eta/\rho$ where η is the entropy of the gas, since it satisfies the same inequality (Clausius-Duhem inequality). This identification is validated by evaluating H at equilibrium, when f must have the form indicated in Eq. (25); in such a case $\eta = -RH$ turns out to have the same dependence on ρ and T as the entropy in ordinary thermodynamics.

Let us now briefly examine the problem of solving the Boltzmann equation; because of the nonlinear nature of the collision term $Q(f,f)$, this is a difficult problem. A very particular class of solutions is offered by Maxwellian distributions, Eq. (25), which describe states characterized by the fact that neither heat flux nor stresses other than isotropic pressure are present. If we want to describe more realistic nonequilibrium situations, we have to rely upon approximate methods, typically perturbation techniques. The simplest approach is to write

$$f = f_0(1 + \varepsilon h_1 + \varepsilon^2 h_2 + \cdots) \qquad (29)$$

where f_0 is a Maxwellian and ε is a "small parameter", which may or may not appear in the Boltzmann equation. In the second case, ε will appear in the initial and boundary conditions and the equation for $h = h_1$ will be

$$\frac{\partial h}{\partial t} + \underline{\xi} \cdot \frac{\partial h}{\partial \underline{x}} = Lh \qquad (30)$$

where

$$Lh = \frac{2}{f_0} Q(f_0, f_0 h) \qquad (31)$$

is called the linearized Boltzmann operator. Eq. (30), in turn, is called linearized Boltzmann equation. If one introduces a Hilbert space $\mathcal{H}$ where the scalar product is given by

$$(g, h) = \int f_0 g h \, d\underline{\xi} \qquad (32)$$

then L is a symmetric operator in $\mathcal{H}$:

$$(g, Lh) = (Lg, h) \qquad (33)$$

In addition, L is non-negative

$$(h, Lh) \leq 0 \qquad (34)$$

and the equality sign holds if and only if h is a collision invariant. In such a case

$$Lh = 0 \tag{35}$$

i.e. the collision invariants are eigenfunctions associated with the fivefold degenerate eigenvalue $\lambda = 0$ of the operator L. All these properties follow immediately from Eq. (31) and (20), if the circumstance that f_0 satisfies Eq. (22) is properly taken into account.

Eq. (35) suggests investigating the spectrum of L; this problem arises when we look for the solution of Eq. (30) in the space homogeneous case $(\partial h/\partial \underline{x} = 0)$. Eq. (34) shows that the spectrum is contained in the negative real semiaxis of the λ-plane; it turns out that the spectrum is extremely dependent upon the form of the choice of the function $B(\theta, V)$ appearing in Eq. (3). It is completely discrete for the case of Maxwell molecules, while it is partly discrete and partly continuous in the case of rigid spheres. For further details, one should consult Refs. [1,2].

An interesting problem arises when one investigates the solutions which do not depend on time t and two space coordinates, say x_2 and x_3 ; in this case one has to solve the equation

$$\xi_1 \frac{\partial h}{\partial x_1} = Lh \tag{36}$$

in the unknown $h = h(x_1, \xi_1, \xi_2, \xi_3) = h(x_1, \underline{\xi})$. The similarity between this equation and Eq. (30) with $\partial h/\partial \underline{x} = 0$ suggests that we look for solutions of the form

$$h = e^{\lambda x_1} g(\underline{\xi}) \tag{37}$$

where g satisfies

$$Lg = \lambda \xi_1 g \tag{38}$$

which is the analogue of $Lh=\lambda h$. The first question is whether the solutions of Eq. (38) are sufficient to construct the general solution of Eq. (36) by superposition. Next comes a study of the set of values of λ for which Eq. (38) has a solution (different from $g = 0$).

The problem here is more difficult because there is an interplay between L and the multiplicative operator ξ_1. In addition the existence of the collision invariants satisfying Eq. (35) prevents L from being a strictly negative operator. In spite of this, it is possible to show [2] that the general solution of Eq. (36) can be written as follows:

$$h = \sum_{\alpha=0}^{4} A_\alpha \psi_\alpha + \sum_{\alpha=2}^{4} B_\alpha \left[L^{-1}(\xi_1 \psi_\alpha) + x_1 \psi_\alpha\right] + \left[\int_{-\lambda_\infty}^{-\lambda_0} + \int_{\lambda_0}^{\lambda_\infty}\right] g_\lambda(\xi) \cdot A_\lambda e^{\lambda x_1} d\lambda . \qquad (A_\alpha, B_\alpha = \text{const.}; A_\lambda = A(\lambda)) \tag{39}$$

where g_λ are the eigensolutions of Eq. (38), ψ_α are the five collision invariants $\psi_0 = 1$, $\psi_i = \xi_i$ $(i = 1,2,3)$, $\psi_4 = \xi^2 - 5RT_0$ (T_0 being the temperature in the basic Maxwellian f_0). Here we have assumed that the λ's form a continuous set, otherwise the corresponding integral has to be replaced by the sum $\sum_k e^{\lambda_k x_1} g_k(\xi) A_k$.

It is clear that the general solution given in Eq. (39) is made up of two parts, h_A and h_B, given by

$$h_A = \sum_{\alpha=0}^{4} A_\alpha \psi_\alpha + \sum_{\alpha=2}^{4} B_\alpha \left[x_1 \psi_\alpha + L^{-1}(\xi_1 \psi_\alpha)\right] \tag{40}$$

$$h_B = \int_{-\lambda_\infty}^{-\lambda_0} + \int_{\lambda_0}^{\lambda_\infty} e^{\lambda x_1} g_\lambda(\xi) A_\lambda d\lambda \tag{41}$$

where the "eigenvalues" λ are of the order of the inverse of the mean free path [2] . It is clear that h_B describes space transients which are of importance in the neighbourhood of boundaries and become negligible a few mean free paths far from them. The circumstance that Eq. (41) contains exponentials with both $\lambda > 0$ and $\lambda < 0$ is exactly what is required to describe a decay either for $x > \bar{x}_1$ or $x_1 < \bar{x}_1$, where $\bar{x}_1$ is the location of a boundary.

The general solution given by Eq. (39) then shows that, if the region where the gas is contained (either a half space or a slab of thickness d, because of the assumption that h is independent of two space coordinates) is much thicker than the mean free path ℓ, then h_B will be negligible except in boundary layers a few mean free paths thick. These layers receive the name of "Knudsen layers" or "Kinetic boundary layers". Outside them the solution is accurately described by the asymptotic part h_A , defined by Eq. (40); it can be shown [2] that if we compute the stress tensor and heat flux vectors arising from h_A , they turn out to be related to the velocity and temperature gradients by the Navier-Stokes-Fourier relations ,with the following expressions for the viscosity coefficient μ and the heat conduction coefficient k :

$$\mu = -(RT_0)^{-1} \int \xi_1 \xi_2 L^{-1}(\xi_1 \xi_2) d\xi \tag{42}$$

$$\kappa = (4RT_0^2)^{-1} \int f_0 \xi_1 \xi^2 L^{-1}[\xi_1 (\xi^2 - 5RT_0)] d\xi \tag{43}$$

These results can be extended to more general problems [2]. Very interesting problems arise when the inequality $d \gg \ell$ is not satisfied, i.e. the mean free path is comparable with the slab thickness [1, 2] ; their treatment is, however, beyond the limits of the present seminar.

REFERENCES

1 - C. Cercignani - "Mathematical Methods in Kinetic Theory", Plenum Press, N.Y. (1969)

2 - C. Cercignani - "Theory and Application of the Boltzmann Equation", Scottish Academic Press, Edinburgh (1975).

CENTRO INTERNAZIONALE MATEMATICO ESTIVO

(C.I.M.E.)

QUALITATIVE AND STATISTICAL THEORY

OF DISSIPATIVE SYSTEMS

Oscar E. LANFORD III

Department of Mathematics, University of California

Berkeley, California 94720

Corso tenuto a Bressanone dal 21 al 24 giugno 1976

Qualitative and Statistical Theory of Dissipative Systems

Oscar E. Lanford III
Department of Mathematics
University of California
Berkeley, California 94720

Preparation of these notes was supported in part by NSF Grant MCS 75-05576. A01.

This series of lectures will be concerned with the statistical theory of dissipative systems and, at least metaphorically, with its applications to hydrodynamics. The principal objective will be to try to clarify the question of how to construct the appropriate ensemble for the statistical theory of turbulence. We will not, however, come to this point for some time. It should be noted at the outset that the relevance of our discussion to the theory of turbulence is dependent on the guess that, despite the fact that fluid flow problems have infinite-dimensional state spaces, the important phenomena are essentially finite dimensional.* This point of view is not universally accepted [4]. On the other hand, the theory is not restricted to fluid flow problems; it also applies to a large number of model systems arising, for example, in mathematical biology [7].

The methods we will discuss are limited in that they appear not to have anything to say about such traditionally central issues as the characteristic spatial properties of turbulent flow, the dynamics of vorticity, etc. Instead, they attempt to clarify the apparently stochastic character of the flow and its peculiar dependence-independence on initial conditions. To explain what this means, let us look briefly at two important but not completely precise distinctions — between conservative and dissipative systems and between stable and unstable ones. Intuitively, when we say that a system is conservative, we mean that, once it has been started in motion, it will keep going forever without

---

*It may be that this ceases to be true for "fully developed turbulence" and that what we say here applies to turbulence at relatively low Reynolds numbers and not at high Reynolds numbers.

further external driving. Mathematically this is usually reflected in the fact that the equations of motion may be written in Hamiltonian form, with the consequent conservation of energy and phase space volume. Among numerous examples, let us note

a. the Newtonian two-body problem

b. the motion of a finite number of frictionless and perfectly elastic billiard balls on a rectangular table.

These examples illustrate the distinction between stable and unstable systems. The Newtonian two-body system is stable in the sense that the effects of small perturbations of the initial conditions grow slowly if at all and hence that long-term predictions about the state of the system are possible on the basis of approximate information about the initial state. In the billiard system, on the other hand, even very small changes in the initial state are soon amplified so that they have large effects. If the system is started out repeatedly, in almost but not exactly the same way, the long-term histories will almost certainly be totally different. In this sense, although the motion is strictly speaking deterministic, it is from a practical point of view effectively random; the coarse features of the state of the system at large times depend on unobservably fine details of the state at time zero.

Consider next dissipative systems. Intuitively, these have some sort of frictional mechanism which tends to damp out motion and must therefore be driven by external forces if they are not simply to stop. A mathematical transcription of this notion which is as general as the correspondence "conservative ≡ Hamiltonian" does not seem to exist, but it is generally not difficult to agree on whether a given dynamical system is dissipative or not. We will consider systems driven by time-independent forces, such as a viscous fluid flowing through a pipe or electric circuits driven by batteries. In many cases these systems display behavior which is simpler than that of conservative systems -- they may tend, independent of how they are started out, to a dynamical equilibrium in which driving forces are exactly balanced by dissipation. A system which tends to the same equilibrium, no matter where in its state space it starts, appears to forget its initial conditions and hence to be "even more stable" than the conservative Newtonian two-body system considered above. Long-term predictions can be made which don't depend on the initial state but only on the parameters appearing in the equations of motion.

The next simplest possible behavior is the existence of a globally attracting periodic solution or limit cycle. In this case the equations of motion admit a solution $x_0(t)$ with $x_0(\tau) = x_0(0)$ for some $\tau > 0$, and every solution of the equations of motion converges to the set $\{x_0(t): 0 \leq t \leq \tau\}$ as $t \to \infty$. What usually happens in this situation is in fact something more special: For each initial x_1 there exists $t_1(x_1)$ with $0 \leq t_1 < \tau$ such that

$$\lim_{t\to\infty} d(x_1(t), x_0(t + t_1)) = 0 .$$

Although the long-time behavior is no longer completely independent of the initial point, the role of the initial point is simply to determine the phase t_1. Again the motion satisfies our intuitive criterion for stability; the long-term effect of a small change in the initial point is simply a small change in the phase.

It is natural to ask what comes next after periodic orbits in the hierarchy of complexity for dissipative systems. One plausible guess, advocated by Landau among others, is that instead of having a single period, the system may have two or more independent periods -- i.e., that the state space may contain a torus of dimension two or greater which is invariant under the solution flow, which attracts at least nearby solution curves, and on which the solution flow reduces in appropriate co-ordinates to uniform velocity flow. Although this certainly can happen, it is not likely to be common since it is destroyed by most small perturbations when it does occur. What turns out to be much more likely is the presence of what have come to be called "strange attractors" sets invariant under the solution flow and attracting nearby orbits but which, instead of being smooth manifolds like tori, have a complicated Cantor-set-like structure. We will present shortly a simple example of a system with such a strange attractor, but before doing so we need to introduce some notions from the qualitative theory of differential equations.

The states of the physical system we are considering will be assumed to form a manifold M which we will take to be finite-dimensional (although much of the formal theory extends easily to infinite-dimensional manifolds). The equations of motion will be taken to be first-order ordinary differential equations on M which we will write in the classical co-ordinate form

$$\frac{dx_i}{dt} = F_i(x_1,\ldots,x_n) \quad 1 \leq i \leq n,$$

where n is the dimension of M. To avoid uninteresting complications we will assume that the right-hand side is an infinitely differentiable function of $x_1,\ldots,x_n$. We will denote the solution mappings by T^t, so $T^t x$ is the solution curve passing through x at time zero. We will assume that, for any x, $T^t x$ exists for all $t > 0$ and $\{T^t x: t > 0\}$ is relatively compact. There are many interesting cases in which this condition is satisfied but in which solution curves do not in general exist for all $t < 0$; the condition is automatically satisfied if the state space M is compact.

The mathematical transcription of the existence of a dynamic equilibrium to which the system tends no matter how it is started out is as follows: There exists a stationary solution x_0 of the equations of motion such that

$$\lim_{t\to\infty} T^t x = x_0$$

for all $x \in M$. Such an x_0 is said to be a <u>globally attracting</u> stationary solution. More generally, a stationary solution x_0 is <u>locally attracting</u> if

$$\lim_{t\to\infty} T^t x = x_0$$

for all x in some neighborhood of x_0. While it is generally difficult to determine whether a stationary solution is globally attracting, there is a simple sufficient condition for a stationary solution to be a local attractor: It suffices that the matrix of partial derivatives

$$(DF(x_0))_{ij} = \left.\frac{\partial F_i}{\partial x_j}\right|_{x = x_0}$$

giving the linearized equation of motion at x_0 have all its eigenvalues in the open left half-plane.

We have already defined what we mean by saying that a periodic solution to the equations of motion is globally attracting; we will similarly say that a periodic solution $(\bar{x}(t))_{0\leq t\leq \tau}$ with period τ is <u>locally attracting</u> if for all x in some neighborhood of the points set $\{\bar{x}(t): 0 \leq t \leq \tau\}$ we have

$$\lim_{t_1\to\infty} d(T^{t_1}x, \{\bar{x}(t): 0 \leq t \leq \tau\}) = 0 .$$

To give a linear criterion for periodic solution to be locally attracting which is analogous to the one given above for stationary solutions, we introduce the notion of the <u>Poincaré map</u> associated with a periodic solution. Take a small piece Σ of $n-1$ dimensional surface transverse to the periodic solution. For each y on Σ and sufficiently near to the periodic solution define $\Phi(y)$ to be the first point on the forward solution curve $\{T^t y: t > 0\}$ which is again in Σ. If y_0 denotes the point where the periodic solution crosses Σ, then $\Phi(y_0) = y_0$. In order that the periodic solution be locally attracting it is sufficient that the derivative matrix

$$\left.\frac{\partial \Phi_i}{\partial y_j}\right|_{y = y_0} \qquad 1 \leq i, j \leq n - 1$$

have all its eigenvalues in the open unit disk. (Here, we have chosen some set $y_1, \ldots, y_{n-1}$ of local co-ordinates for Σ and expressed Φ in terms of these co-ordinates.)

The two simple situations described above — attracting stationary solutions and attracting periodic solutions — turn out to be closely related via the Hopf bifurcation. Suppose that our differential equation depends on a parameter r which may indicate, for example, how hard the system is being driven. Suppose also that for some value r_c of r a stationary solution $x = x_r$ changes from stable to unstable by having a complex conjugate pair of eigenvalues for the linearization

$$\left.\frac{\partial F_i}{\partial x_j}\right|_{x = x_r}$$

at x_r cross from the left to the right half-plane at non-zero speed. It turns out that, under these circumstances, if a certain complicated combination of the first, second, and third partial derivatives of F with respect to x at $x = x_r$, $r = r_c$, is positive, then for r slightly larger than r_c there is an attracting periodic solution which can be regarded as making a small circle around the now-unstable stationary solution x_r. As r decreases to r_c the periodic orbit shrinks down to the single point x_{r_c}. In this case we say that the stationary solution undergoes a normal Hopf bifurcation to a periodic solution. It is also possible for the above-mentioned complicated combination of partial derivatives of F to be negative. In this case no attracting periodic solution is formed. Instead, for r slightly smaller than r_c, there exists an unstable periodic solution near x_r which shrinks down to x_{r_c} as r increases to r_c. In this case we say that the stationary solution undergoes an inverted bifurcation. (Other, more complicated, things can happen if the combination of partial derivatives is zero, or if more than two eigenvalues cross the imaginary axis simultaneously.) For a detailed discussion of the Hopf bifurcation and related phenomena, see [6].

We next need some more general notions which apply even in the absence of stationary and periodic solutions. For any point x in the state space we define the ω-limit set of x, denoted by ω(x), to be

$$\omega(x) = \bigcap_{\tau\to\infty} \overline{\{T^t x\colon t \geq \tau\}}.$$

Alternatively, ω(x) is the set of all cluster points of sequences $T^{t_n}x$ with $t_n \to \infty$. The assumption that $\{T^t x\colon t \geq 0\}$ is relatively compact implies that ω(x) is not empty; ω(x) is also evidently closed and invariant under the solution flow T^t. If the solution curve $T^t x$ converges as $t \to \infty$ to a stationary solution x_0 then $\omega(x) = \{x_0\}$; conversely, if ω(x) contains only one point x_0 then x_0 is a stationary solution and $\lim_{t\to\infty} T^t x = x_0$. Similar statements hold for solution curves converging to a periodic solution.

Let

$$\tilde{\Omega} = \bigcup_x \omega(x).$$

Then every forward solution curve $T^t x$ converges as $t \to \infty$ to $\tilde{\Omega}$, and

$\tilde{\Omega}$ is the smallest closed set with this property. In order to understand the behavior of solution curves for large positive times, it is enough to study the solution flow on and near $\tilde{\Omega}$, i.e., $\tilde{\Omega}$ is the essential part of the state space from the long-term point of view. One important difference between dissipative and non-dissipative systems is that $\tilde{\Omega}$ tends to be small for dissipative systems and to be the whole state space for conservative systems. To see the latter fact we first note:

Proposition. *Any finite measure* μ *invariant under the solution flow* T^t *has support in* $\tilde{\Omega}$.

Proof. We want to show that $\int \varphi d\mu = 0$ for any continuous function whose support is compact and disjoint from $\tilde{\Omega}$. By the assumed invariance of μ under T^t,

$$\int \varphi \, d\mu = \int \varphi \circ T^t \, d\mu \quad \text{for any } t.$$

Now, since the support of φ is disjoint from $\tilde{\Omega}$,

$$\lim_{t\to\infty} \varphi(T^t x) = 0 \quad \text{for all } x$$

so by the dominated convergence theorem

$$\int \varphi d\mu = \lim_{t\to\infty} \int \varphi \circ T^t d\mu = \int \lim_{t\to\infty} \varphi \circ T^t \, d\mu = 0.$$

If, for example, we consider a Hamiltonian system with the property that $\{x: H(x) \leq E\}$ is compact for each E, then Licuville's Theorem implies that every point of the state space is in the support of some invariant measure and hence, by the proposition, that $\tilde{\Omega}$ is the whole state space.

We will say that a point x of the state space is a wandering point for T^t if there is a neighborhood U of x such that

$$T^t U \cap U = \emptyset \text{ for all sufficiently large } t.$$

The non-wandering set Ω is the set of all points which are not wandering. It follows at once from the definition that the set of wandering points is open and hence that Ω is closed; it is also easy to see that

$$\omega(x) \subset \Omega \qquad \text{for all } x.$$

Hence, $\tilde{\Omega} \subset \Omega$. While it is not difficult to construct flows for which $\tilde{\Omega} \neq \Omega$, equality does hold in most interesting cases (including flows which satisfy Smale's Axiom A, to be described below). If the solution flow has a globally attracting stationary solution x_0, then

$$\tilde{\Omega} = \Omega = \{x_0\} ;$$

a similar statement holds when there is a globally attracting periodic solution.

The simplest asymptotic behavior a differential equation can have is for all solution curves to converge to the same stationary or periodic solution. One trivial way in which the situation can become more complicated is to have several locally attracting stationary and periodic solutions. If x_0 is a locally attracting stationary solution define the _basin of attraction of x_0_ to be

$$B(\{x_0\}) = \{x: \lim_{t\to\infty} T^t x = x_0\} .$$

One can define in a similar way the basin of attraction of a periodic solution. A basin of attraction is open and invariant under the solution flow. (The fact that it is open may not be quite obvious. Consider the case of a locally attracting stationary solution x_0 and let x be in the basin of attraction of x_0. Because x_0 is locally attracting there is a neighborhood V of x_0 such that every solution curve beginning in V converges to x_0, and because $T^t x$ converges to x_0 there exists t_0 such that $T^{t_0} x \in V$. Then $(T^{t_0})^{-1} V$ is an open neighborhood of x contained in the basin of attraction of x_0.) Note that, since basins of attraction are open and disjoint, the state space M, if connected, cannot be written as the union of two or more basins. Hence, if there are at least two locally attracting stationary or periodic solution, there must be some solution curves (lying on the boundaries of the basins of attraction) which do not converge to any of them.

We are going to investigate attractors which are more complicated than single points and periodic solutions, and we should therefore define precisely and generally what we mean by an attractor. Unfortunately, no such definition seems to be agreed upon, so we will improvise by listing a number of properties which an attractor ought to have, being careful that the conditions are indeed satisfied in the special case of Axiom A attractors, where there does exist an accepted definition (to be discussed below). To begin with, an attractor should be a closed (compact?) subset X of the state space, invariant under the solution flow, which attracts nearby orbits in the sense that there exists an open set U containing X such that, for any x in U

$$\lim_{t\to\infty} d(T^t x, X) = 0$$

(or equivalently, $\omega(x) \subset X$). Second, we require that if x is near X then $T^t x$ remains near X for all $t > 0$. The <u>basin of attraction</u> $B(X)$ of X is now defined to be

$$\{x \in M: \omega(x) \subset X\} .$$

The argument given above for attracting stationary solutions is easily adapted to show that $B(X)$ is open. We also want to put into the definition some condition which prevents an attractor from being decomposable into a finite number of other attractors; a good way to do this is to require that some solution curve contained in the attractor is dense in the attractor, i.e., that the solution flow T^t restricted to X is <u>topologically transitive</u>. Nothing in the above list of conditions prevents the whole state space from being an attractor; it will be one if there is a single solution curve dense in the whole state space, and this frequently happens for conservative systems (if the state space is taken to be a single energy surface). For dissipative systems, on the other hand, attractors will generally be small at least in the sense of having empty interiors.

The preceding discussion has considered only continuous flows T^t.

For many purposes it is useful to have a parallel set of definitions for the corresponding discrete situation, the set of powers $\{T^n\}$ of a single smooth transformation T of a manifold. The task of adapting our discussion to this slightly different context is straightforward; we leave it to the reader.

Up to this point, we have been dealing with elementary general considerations. Although necessary in order to get started, the ideas developed so far do not seem to be sufficiently specific to lead to any very interesting analysis. In order to go further we must impose additional restrictions on the systems we study. In recent years, it has turned out to be particularly fruitful to impose some sort of <u>hyperbolicity</u> condition. The fundamental reference in this area is [11]; we will sketch here a few of the basic ideas.

We explained above that a stationary solution x_0 is at least a local attractor if all eigenvalues of the matrix

$$(DF)_{ij} = \frac{\partial F_i}{\partial x_j}\bigg|_{x = x_0}$$

are in the open left half-plane. More generally, we say that x_0 is a <u>hyperbolic</u> stationary solution if no eigenvalues for DF are precisely on the imaginary axis. In this case. $\mathbb{R}^n$ splits into two complementary subspaces E^s and E^u, each invariant under DF, such that the eigenvalues of $DF|_{E^s}$ have strictly negative real parts while the eigenvalues of $DF|_{E^n}$ have strictly positive real parts; E^s and E^u are called respectively the stable and unstable eigenspaces for DF. If we let $\tilde{F}$ denote the linearization of F at x_0, i.e.

$$\tilde{F}(x) = DF(x - x_0)$$

then a solution $\tilde{x}(t)$ of the linearized equation

$$\frac{d\tilde{x}}{dt} = \tilde{F}(\tilde{x})$$

converges to x_0 as $t \to \infty$ if and only if $\tilde{x}(0) \in x_0 + E^s$ and converges to x_0 at $t \to -\infty$ if and only if $\tilde{x}(0) \in x_0 + E^u$. Going back to the full (non-linear) equation we define the <u>stable</u> and <u>unstable</u> manifolds at x_0 ($W^s(x_0)$ and $W^u(x_0)$ respectively) by

$$W^s(x_0) = \{x \in M: T^t x \to x_0 \text{ as } t \to \infty\}$$

$$W^u(x_0) = \{x \in M: T^t x \to x_0 \text{ as } t \to -\infty\}$$

From the definition it is not apparent that these sets are submanifolds, or even that they contain any points other than x_0, but we have:

<u>Stable Manifold Theorem for Hyperbolic Fixed Points.</u> $W^s(x_0)$ *and* $W^u(x_0)$ *are submanifolds* of* M, *with dimensions equal respectively to dim* $E^s(x_0)$ *and dim* $E^u(x_0)$. *These submanifolds contain* x_0 *and are tangent at* x_0 *to* $E^s(x_0)$ *and* $E^u(x_0)$ *respectively.*

* There is a technical distinction which needs to be noted here. The stable and unstable manifolds are <u>immersed</u> but not in general <u>imbedded</u> submanifolds of M. This means that, although made up of countably many smooth pieces, they can fold back arbitrarily near themselves. A simple example of an immersed one-dimensional submanifold of $\mathbb{R}^3$ is the "Lissajous figure"

$$\{(\cos(t), \cos(\omega_1 t), \cos(\omega_2 t)): -\infty < t < \infty\}$$

where ω_1, ω_2, and ω_1/ω_2 are all irrational.

With these elementary examples for motivation, we will now give a general definition of hyperbolic set. There are in fact two definitions, one for transformations Φ and one for flows T^t. We will give only the definition for transformations. Thus, let Φ be a differentiable and invertible mapping and let Λ be a compact set mapped onto itself by Λ. For the sake of concreteness we will assume that a single set of co-ordinates can be chosen for an open set containing Λ (i.e., we will act as if Λ is contained in $\mathbb{R}^n$); there is, however, no difficulty in eliminating this assumption by giving a coordinate free version of the definition. We define the derivative of Φ at x to be the $n \times n$ matrix

$$(D\Phi)_{ij}(x) = \frac{\partial}{\partial x_j} \Phi_i(x)$$

and similarly define $D\Phi^m(x)$ for any integer m (positive or negative). By the chain rule

$$D\Phi^{m+p}(x) = D\Phi^m(\Phi^p x) \circ D\Phi^p(x).$$

We are going to define Λ to be a hyperbolic set for Φ if, roughly speaking, any infinitesimal displacement from a point x belonging to Λ can be decomposed as the sum of two infinitesimal displacements, one of which contracts exponentially under positive powers of Φ and the other of which contracts exponentially under negative powers of Φ. More precisely: Λ is a <u>hyperbolic set</u> for Φ if there exists for each $x \in \Lambda$ a splitting of $\mathbb{R}^n$ into a direct sum of complementary subspaces $E^s(x)$, $E^u(x)$ such that:

For some $c > 0$, $\lambda < 1$, which do not depend on x or m,

$$\left\| D\Phi^m(x)\big|_{E^s(x)} \right\| \leq c\lambda^m \qquad m = 1,2,3,\ldots$$

$$\left\| D\Phi^{-m}(x)\big|_{E^u(x)} \right\| \leq c\lambda^m$$

In addition the splitting is required to be invariant under Φ:

$$D\Phi E^s(x) = E^s(\Phi(x))$$

$$D\Phi E^u(x) = E^u(\Phi(x))$$

and to vary continuously with x. This last condition means the following: For every $x_0 \in \Lambda$, there exists an open neighborhood U of x_0 such that $n^s = \dim E^s(x)$ and $n^u = \dim E^u(x)$ are constant on U and such that there exist n continuous $\mathbb{R}^n$-valued functions $\xi_1(x),\ldots,\xi_n(x)$ defined on $U \cap \Lambda$ such that, for each $x \in \Lambda \cap U$, $E^s(x)$ is spanned by $\xi_1(x),\ldots,\xi_{n^s}(x)$ and $E^u(x)$ is spanned by $\xi_{n^s+1}(x),\ldots,\xi_n(x)$. Alternatively, as it turns out, it is enough to require that

$$\{(x,\xi)\colon \xi \in E^s(x)\} \quad \text{and} \quad \{(x,\xi)\colon \xi \in E^u(x)\}$$

are both closed subsets of $\Lambda \times \mathbb{R}^n$; continuity as formulated above then follows automatically.

It is particularly interesting to apply this definition with Λ the non-wandering set of Φ. We say that Φ satisfies Axiom A if

1. Ω is a hyperbolic set

2. The periodic points for Φ are dense in Ω.

This condition has proved very fruitful for mathematical analysis. It is, on the other hand, hard to verify in practical applications and non-trivial examples are relatively scarce. It is at this time still too early to decide whether Axiom A as it stands is too restrictive to apply to cases of interest, but either it or some weakened version of it seems likely to play an important role in future developments.

It may be helpful to note here one difference between hyperbolic fixed points and more general hyperbolic sets. A hyperbolic fixed point x cannot be an attractor unless $E^u(s)$ is trivial. There is no such restriction for general hyperbolic sets. It frequently happens that a hyperbolic attractor is made up locally of infinitely many smooth "leaves" -- lower-dimensional surfaces which are everywhere tangent to $E^u(x)$. Two nearby points on the same leaf move apart under the action of the transformation, but the whole assembly of leaves is attracting.

Chapter II. The Lorenz System

We turn now from generalities to a discussion of a particular system of equations. This system could hardly be simpler -- the state space is three-dimensional and the equations are

$$\frac{dX}{dt} = -\sigma X + \sigma Y$$

$$\frac{dY}{dt} = rX - Y - XZ$$

$$\frac{dZ}{dt} = XY - bZ$$

with b, σ, r positive constants -- but it displays a bewildering assortment of non-trivial mathematical phenomena. So far as I know, this system of equations was first seriously investigated by E. N. Lorenz [5] some fifteeen years ago; in recent times it has been studied intensively by Yorke, Guckenheimer [3], Martin and McLaughlin [8], Marsden and McCracken [6], and Williams [12], among others.

One of the appealing aspects of the Lorenz system is the fact that it was not constructed for the purpose of proving the possibility of complicated behavior; rather, it turned up in the course of a practical investigation. In his original paper, Lorenz was led to this system by the following considerations: Consider the equations of motion for two-dimensional convection in a container of height H and length L. These equations can be viewed, heuristically at least, as a first order differential equation on an infinite dimensional state space; the points of the state space are pairs consisting of a stream function $\psi(x,z)$ and a temperature field $T(x,z)$ subject to appropriate boundary conditions. Look for solutions of the form

$$\psi(t,x,z) = \sum_{n=1}^{\infty} \sum_{m=1}^{\infty} \psi_{m,n}(t) \sin(\tfrac{\pi x}{L}) \sin(\tfrac{\pi z}{H})$$

$$T(t,x,z) = T_- + (\tfrac{z}{H})(T_+ - T_-) + \sum_{n=1}^{\infty} \sum_{m=0}^{\infty} \theta_{m,n}(t) \cos(\tfrac{\pi x}{L}) \sin(\tfrac{\pi z}{H}),$$

where T_-, T_+ denote the temperatures at the bottom and the top of the container respectively. (Such solutions correspond to free or no-stress boundary conditions on the velocity field and to the absence of heat flow through the ends of the container.) Express the equations of motion directly in terms of the Fourier coefficients $\theta_{m,n}(t)$, $\psi_{m,n}(t)$ and drastically truncate the resulting infinite set of coupled differential equations by putting all $\psi_{m,n}, \theta_{m,n}$ identically equal to zero except $\psi_{1,1}, \theta_{1,1}, \theta_{0,2}$. Now put

$$X = c_1\psi_{1,1} \; ; \; Y = c_2\theta_{1,1} \; ; \; Z = c_3\theta_{0,2}$$

and choose c_1, c_2, c_3 so as to simplify the differential equation; the result is the Lorenz system.

With this derivation comes a physical interpretation for the parameters b,σ,r. Specifically, b is a simple geometric constant $(4/(1 + (H/L)^2))$, σ is the Prandtl number (i.e., the ratio of viscosity to thermal conductivity) and r is a numerical constant times the Rayleigh number, i.e., is a dimensionless number proportional to the temperature difference $(T_- - T_+)$. The values of b and σ will be fixed at 8/3 and 10 respectively. We will first discuss schematically how the behavior of typical solutions changes with r; then describe in detail what happens for a particular value of r.

We begin with a number of elementary observations about the Lorenz system:

i) The equations are invariant under the transformation

$$X \to -X; \quad Y \to -Y; \quad Z \to Z$$

(The physical origin of this symmetry is invariance of the equations of motion under reflection through a vertical line at the center of the container.)

ii) The solution flow T^t generated by the Lorenz system shrinks volumes in the state space $\mathbb{R}^3$ at a uniform rate. This follows from the equation

$$\frac{\partial \dot{X}}{\partial X} + \frac{\partial \dot{Y}}{\partial Y} + \frac{\partial \dot{Z}}{\partial Z} = -(\sigma + b + 1) = -13\ 2/3\ .$$

This rate is in fact quite large; a set of states occuping unit volume at time zero occupies only the volume $e^{-13.67} \simeq 10^{-6}$ at time one.

iii) All solutions are bounded for $t > 0$, and very large initial values of X, Y, Z are damped by the motion. To show this we introduce

$$u(t) = (X(t))^2 + (Y(t))^2 + (Z(t) - (r + \sigma))^2.$$

An elementary computation gives

$$\frac{du}{dt} \leq - c_1 u + c_2$$

with constants c_1, c_2 which do not depend on X,Y,Z, but may depend on r,σ,b. (The essential point is that, despite the quadratic terms in the equations of motion, there are no cubic terms in $\frac{du}{dt}$.) It follows easily that every solution curve eventually gets and stays in the interior of the ball B where $u \leq 2\, c_2/c_1$. This ball is mapped into itself by T^t, and by ii) the volume of its image under T^t goes to zero as t goes to infinity. Hence, every solution curve $T^t\underline{x}$ converges to large t to the set

$$\bigcap_{t \geq 0} T^t B$$

which is closed and has Lebesgue measure zero. The non-wandering set Ω for T^t is contained in this intersection and therefore also has measure zero.

We now describe what happens as r is varied starting from zero. Recall that, in the derivation from the convection equations, r was proportional to the imposed vertical temperature difference and is therefore a measure of how hard the system is being driven. For r between zero and one, inclusive, it is not hard to show that $\underline{0}$ is a globally attracting stationary solution. As r is increased past one, this solution becomes unstable and bifurcates into a pair of locally attracting stationary solutions $\underline{C} = (\sqrt{b(r-1)}, \sqrt{b(r-1)}, r-1)$,

$\underline{C}' = (-\sqrt{b(r-1)}, -\sqrt{b(r-1)}, r-1)$. These are easily checked to be the only stationary solutions aside from $\underline{0}$; they remain present, but not necessarily stable, for all $r > 1$. Physically, they represent steady convection. Also for all $r > 1$ the stationary solution $\underline{0}$ is hyperbolic, with a two-dimensional stable manifold and a one-dimensional unstable manifold. For r slightly greater than one, nearly all solution curves converge either to $\underline{C}$ or to $\underline{C}'$ for large time; the only exceptions are those on the two-dimensional stable manifold of $\underline{0}$.

The two steady convection solutions $\underline{C}$ and $\underline{C}'$ remain stable until $r = 470/19 \approx 24.74$ (for $\sigma = 10$), but various interesting things happen before then. At a special value of r around $r = 13.9$ the one dimensional unstable manifold for $\underline{0}$ returns to $\underline{0}$ (homoclinic orbit), and for slightly larger values of r there are two unstable hyperbolic periodic orbits. It is not known whether the non-wandering set Ω remains equal to $\{\underline{C}, \underline{C}', \underline{0}\}$ all the way up to this value of r or whether periodic solutions or other kinds of recurrence appear earlier. Nevertheless, the appearances are that, until r is nearly equal to $470/19$, all solution curves except for a set of measure zero converge to one of $\underline{C}$ or $\underline{C}'$. This does not remain true all the way up to $r = 470/19$, but what happens for r's slightly below that critical value is most easily understood in terms of what happens above it.

As r passes the critical value of $470/19$, both $\underline{C}$ and $\underline{C}'$ become unstable through having a complex-conjugate pair of non-real eigenvalues cross into the right half plane. This does not lead, via a normal Hopf bifurcation, to stable periodic orbits near $\underline{C}$ and $\underline{C}'$

for r slightly above 470/19. Instead (see [8], [6]), what happens is an inverted Hopf bifurcation in which an unstable periodic solution contracts to each of $\underline{C},\underline{C}'$ and disappears as r increases to 470/19. The behavior above the critical value of r seems not to be accessible to analysis by "infinitesimal" bifurcation theory but requires a global investigation of the behavior of solutions which has so far been possible to carry out only by following solutions numerically on a computer. The next stage of our discussion will be a description of the results of such a numerical investigation, carried out for the arbitrarily chosen value $r = 28$.

In the vicinity of any one of the three stationary solution $\underline{0},\underline{C},\underline{C}'$, the motion is similar to that given by the linearization at the stationary solution. For each of the $\underline{C}$ and $\underline{C}'$, the linearization has a pair of complex eigenvalues $.094 \pm 10.2\,i$ and a negative eigenvalue -13.85. Hence, in the linearized motion, three things are going on at quite different speeds:

i) the component in the negative eigendirection damps out rapidly

ii) the component in the two-dimensional real eigenspace associated with the complex eigenvalue pair rotates at a moderate speed and also

iii) expands slowly.

More specifically the "rotation period" is $2\pi/10.2 \approx .62$; the contracting component is multiplied by about 2×10^{-4} for each rotation; and the rotating component expands by about 6% with each rotation. The same qualitative picture holds for the correct (not linearized) motion near $\underline{C}$ and $\underline{C}'$. Passing through each of these points is a two dimensional surface, its unstable manifold, which strongly attracts nearby solution curves and along which solutions spiral slowly outward. The approximate appearance of a typical solution curve is shown in figure 1.

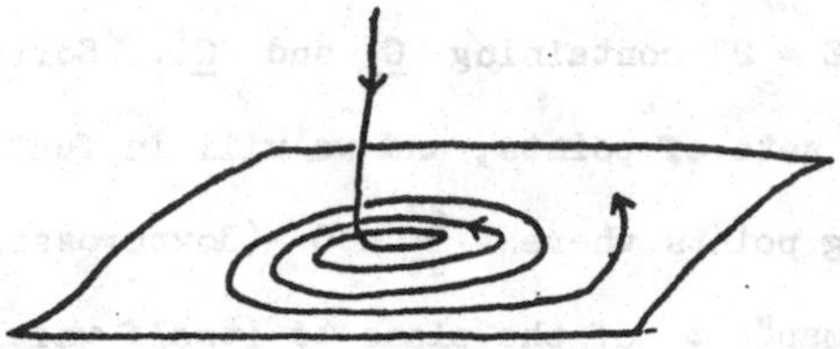

Figure 1

The normal to this surface at $\underline{C}$ has polar angle 70° with respect to the Z axis and azimuthal angle 153° with respect to the X axis.

As it turns out, there is quite a large domain around each of $\underline{C}$, $\underline{C}'$ where this picture is qualitatively correct. This does not yet tell us much about the asymptotic behavior of typical solutions since the steady growth of the rotating component eventually drives the solution curve out of this domain, and we have to look at where it goes. The key to understanding the recurrent behavior of the Lorenz system is the fact that it usually goes into the corresponding domain around the other steady convection solution, where it is again attracted to the unstable manifold, eventually pushed out again, returns to the original domain, and proceeds to repeat the whole process. The repetition is, however, typically only approximate and may differ quite a lot from the first cycle in detail. Although most orbits continue forever shuttling back and forth between C and C′, they are only exceptionally periodic or even asymptotic to periodic solutions.

To form a more precise picture of this process we take a section with the horizontal plane $Z = 27$ containing C and C′. Solution curves then become simply discrete sets of points, and we will in fact keep track only of these crossing points where $\frac{dZ}{dt} < 0$ (downcrossings). We thus define a "Poincaré map" Φ of the plane to itself which takes each point to the next downcrossing on its solution curve. (As we shall see, Φ is not defined everywhere, but it does turn out to be defined almost everywhere.) Figure 2 shows what happens to a domain around C when Φ is applied to it a few times; the transverse scale is grossly exaggerated and the strips are really much thinner than indicated:

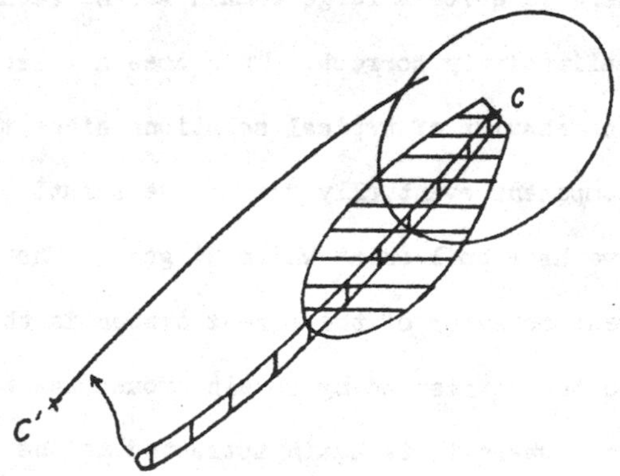

Figure 2

The figure raises the question of what Φ does to the strip running from the domain around $\underline{C}$ to the domain around $\underline{C}'$. Since one end is attached to $\underline{C}$ and the other is attracted to the unstable manifold of $\underline{C}'$, continuity considerations would seem to suggest that the strip will get pulled diagonally across from $\underline{C}$ to $\underline{C}'$. Such is, however, not the case: Although the solution flow is continuous, the Poincaré map Φ need not be, and it in fact undergoes a jump discontinuity between $\underline{C}$ and $\underline{C}'$.

The source of this discontinuity is the third stationary solution at $\underline{0}$. Recall that $\underline{0}$ is a hyperbolic stationary solution with a two-dimensional stable manifold. It is easy to see from the equations of motion that the Z axis is contained in the stable manifold of $\underline{0}$, so we should expect the stable manifold to intersect the plane $Z = 27$ in a curve passing through $X = Y = 0$. This turns out to be correct; we will call the curve in question Σ; it is shown in Figure 5 as running from upper left to lower right. (Actually, this is only one of infinitely many pieces of the intersection of the stable manifold of $\underline{0}$ with the plane $Z = 27$.) Solution curves starting on Σ simply proceed monotonically to $\underline{0}$ and never return to the plane $Z = 27$; Φ is not defined along Σ. Let us investigate what $\Phi(\underline{x})$ does as $\underline{x}$ approaches Σ from the upper right by tracing the orbit $T^t\underline{x}$ for $\underline{x}$ very slightly above Σ. For a long time $T^t\underline{x}$ tracks its neighbors on the stable manifold and hence gets very close to $\underline{0}$. While it is close to $\underline{0}$ its motion is well approximated by the linearization of the equations of motion at zero. The linearization has two negative eigenvalues and one positive one; the eigenvector corresponding to the positive eigenvalue is horizontal (i.e. has Z-component zero). In the linearized motion, the components in the negative eigendirections decay steadily to zero while the component in the positive eigendirection grows. Initially the negative eigencomponents are much larger than the positive one (since the trajectory comes in near the stable manifold), but, unless the positive eigencomponent is exactly zero, it will eventually dominate the others and the trajectory will move away from zero along the positive eigendirection. The modifications introduced into this picture by the non-linear terms in the interaction are simple: A trajectory entering the vicinity of zero near the stable manifold leaves along the unstable manifold; the closer it is to the stable manifold initially, the closer it will get to zero and the closer it will be to the unstable manifold when it leaves. The unstable manifold consists of two solution curves, growing out of the origin in opposite directions; which of these branches will be followed is determined by which side of the stable manifold the trajectory lies on. See figure 3.

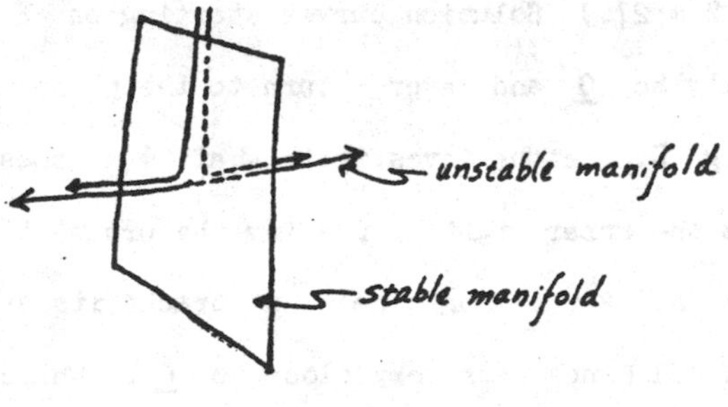

Figure 3.

Since the solution we are following starts slightly above Σ in figure 5, it will eventually be picked up by the branch of the unstable manifold of $\underline{0}$ along which X and Y initially increase. This piece of the unstable manifold makes a large loop around $\underline{C}$, as shown schematically in figure 4, and makes its first downcrossing of the plane Z = 27 at the point $\underline{A}$, with co-ordinates (-5.2, -8.3). (For comparison, the co-ordinates of $\underline{C}'$ are (-8.5, -9.5))

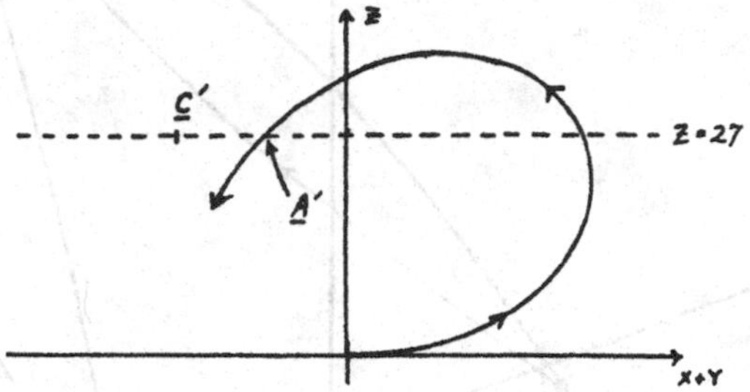

Figure 4

Thus, if $\underline{x}$ is any point on the plane Z = 27 lying near but slightly above Σ, $\Phi(\underline{x})$ is near $\underline{A}$. On the other hand, if $\underline{x}$ is slightly <u>below</u> Σ, $\Phi(x)$ is near $\underline{A}'$, the first downcrossing of the other branch of the unstable manifold of $\underline{0}$ (i.e., the symmetric image of $\underline{A}$ under $X \to -X$, $Y \to -Y$, $Z \to Z$).

The picture, as developed so far, is shown in figure 5. The arc $\underline{C}\,\underline{A}$ is part of the intersection of the unstable manifold of $\underline{c}$ with the plane Z = 27; its symmetric image $\underline{C}'\,\underline{A}'$ is the corresponding object for $\underline{C}'$. We have indicated by arrows the images of a few important points under Φ, and we have put $\underline{B}' = \Phi(\underline{A})$, $\underline{B} = \Phi(\underline{A}')$. This figure, unlike the others in this chapter, is drawn carefully to scale. Note that, although $\underline{B}'$ appears to lie on $\underline{C}'\underline{A}'$, it must in fact be slightly to the right of it.

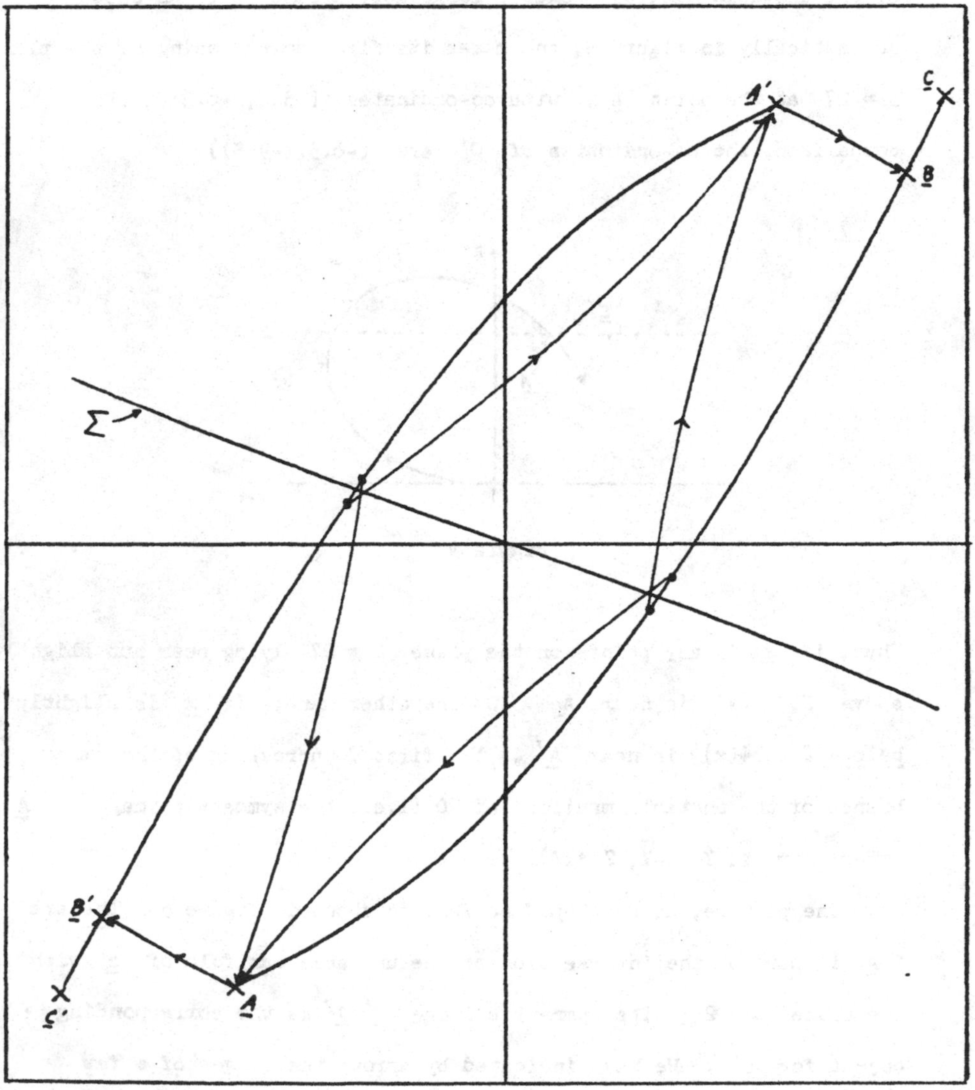

Figure 5

We are now able to form a fairly comprehensive image of the behavior of a typical solution which starts, say, near $\underline{C}$. Such a solution is quickly attracted to the unstable manifold of $\underline{C}$; then proceeds to spiral out along it. Unless it happens to hit Σ exactly, it will eventually land somewhere on the part of $\underline{C}\,\underline{A}$ below Σ. It should then be regarded as spiralling around $\underline{C}'$ rather than $\underline{C}$, and its next downcrossing will be somewhere very near $\underline{C}'\underline{A}'$. Since points near Σ go to the neighborhood of $\underline{A}'$ while $\underline{A}$ goes to $\underline{B}'$, the general point can land anywhere between $\underline{B}'$ and $\underline{A}'$, and what happens next depends on whether it lands above or below Σ. If it lands below Σ, it spirals around $\underline{C}'$ until it is above Σ and then makes a transition back to $\underline{B}\,\underline{A}$; if it lands above Σ it immediately makes the transition back to $\underline{B}\,\underline{A}$. The motion continues in this way forever, alternately moving around $\underline{C}$ and $\underline{C}'$; the only way it can stop is for the solution curve to hit Σ exactly, and this is extremely unlikely.

A few other features of the motion should be noted here. First of all, although the motion may start arbitrarily near to $\underline{C}$, once it has gotten away it can never return to $\underline{C}\,\underline{B}$ (or to $\underline{C}'\,\underline{B}'$.) There are thus gaps of non-zero size between the stationary solutions $\underline{C}$ and $\underline{C}'$ and the region where the orbit is recurrent. This is not the case with the stationary solution at $\underline{0}$; a typical orbit can be expected to approach very close to $\underline{0}$ (but only very infrequently). Second, recall that Φ stretches out the arc from $\underline{A}$ to Σ along the full length of $\underline{B}'\,\underline{A}'$. The arc from $\underline{A}$ to Σ is the image under Φ of the arc from Σ to its preimage under Φ, i.e., of a single cycle under Φ of $\underline{C}'\,\underline{A}$.

Tracing back through a few applications of Φ we find that the arc from $\underline{A}$ to Σ is the image under Φ^n (n about 24, as it happens) of a rather small piece of the arc $\underline{C}\,\underline{A}$ running from $\underline{x}$ to $\Phi(\underline{x})$, with $\underline{x}$ just slightly outside $\underline{B}$. From the linear approximation we know that the distance from $\Phi(\underline{x})$ to $\underline{C}$ is only about 6% greater than the distance from $\underline{x}$ to $\underline{C}$, so an uncertainty of about 6% in the position of a point along $\underline{A}\,\underline{C}$ near $\underline{C}$ leads to complete uncertainty about where its orbit will land on the arc from $\underline{A}$ to Σ and hence to complete uncertainty about where on $\underline{B}'\,\underline{A}'$ it will go next. Thus, although the motion is completely deterministic, it is unstable in the sense that small changes in the initial position are amplified rapidly. This means that the behavior is effectively random; to determine where an orbit will be after making a number of transitions from $\underline{B}\,\underline{A}$ to $\underline{B}'\,\underline{A}'$ and back requires unreasonably precise knowledge about its initial position. Finally, we need to clarify an apparently contradictory aspect of the above description. Points on Σ are on the stable manifold of $\underline{0}$ and the corresponding solution curves approach zero fairly directly. It is natural to visualize the stable manifold of $\underline{0}$ as remaining more or less flat all the way to infinity and hence as separating $\underline{C}$ from $\underline{C}'$. Since a solution curve cannot cross the stable manifold of $\underline{0}$ it would seem to be impossible for solutions to cycle back and forth between $\underline{C}$ and $\underline{C}'$. The fallacy in this argument lies in an incorrect guess about the global structure of the stable manifold of $\underline{0}$. Solution curves passing through the upper left-hand part of Σ, when followed <u>backwards</u>, spiral around $\underline{C}'$ and those near the intersection of $\underline{C}'\,\underline{A}'$ with Σ

spiral around it arbitrarily often. Thus, one part of the stable manifold of $\underline{O}$ wraps infinitely often around $\underline{C}'$. Another part wraps infinitely often around $\underline{C}$, but in the opposite direction. The global structure of the stable manifold of $\underline{O}$ is quite complicated, and it manages to stay out of the way of the motion as described above.

Having seen what a typical solution curve looks like, we will next try to construct a more comprehensive view of all solution curves beginning in a neighborhood of $(\underline{B}\ \underline{A}) \cup (\underline{B}'\ \underline{A}')$. Consider a neighborhood as sketched in Figure 6, consisting of two enlongated ovals S and S':

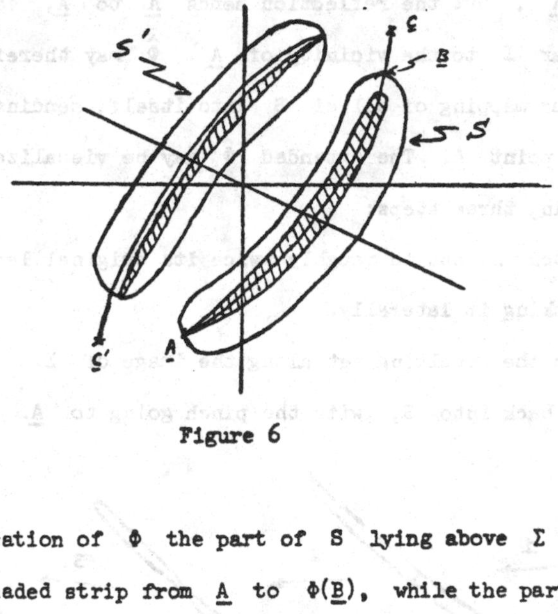

Figure 6

Under one application of Φ the part of S lying above Σ is mapped to the narrow shaded strip from $\underline{A}$ to $\Phi(\underline{B})$, while the part of S lying below Σ is mapped to the narrow shaded strip running from $\underline{B}'$ to $\underline{A}'$ but slightly below $\underline{C}'\ \underline{A}'$. The image of S' is the union of two similar strips. We can at this point simplify the picture substantially by exploiting the symmetry of the Lorenz system. We will identify points in pairs $(-X, -Y, 2T)$ and $(X,Y,2T)$ and represent each such pair by its member in the half-plane $X \geq Y$ (with some appropriate convention on the line $X = Y$); correspondingly, we replace Φ by the quotient mapping $\hat{\Phi}$ obtained by reflecting the image of Φ whenever it lies in the half-plane $Y > X$. This leaves only half of the picture shown in figure 6: The shaded region running from $\underline{B}'$ to $\underline{A}'$ is reflected to run from $\underline{B}$ to $\underline{A}$, and $\hat{\Phi}$ maps S into itself. There is another advantage to making the identification — it permits $\hat{\Phi}$ to be defined on all of S. Points near Σ are mapped by Φ either near $\underline{A}$ or near $\underline{A}'$, but the reflection sends $\underline{A}'$ to $\underline{A}$, so $\hat{\Phi}$ sends all points near Σ to the vicinity of $\underline{A}$. $\hat{\Phi}$ may therefore be extended to a continuous mapping of all of S into itself, sending the arc $\Sigma \cap S$ to the point $\underline{A}$. The extended $\hat{\Phi}$ may be visualized as obtained by the following three steps:

1. Stretch S out to roughly twice its original length, while shrinking it laterally.
2. Pinch the resulting set along the image of Σ.
3. Fold back into S, with the pinch going to $\underline{A}$.

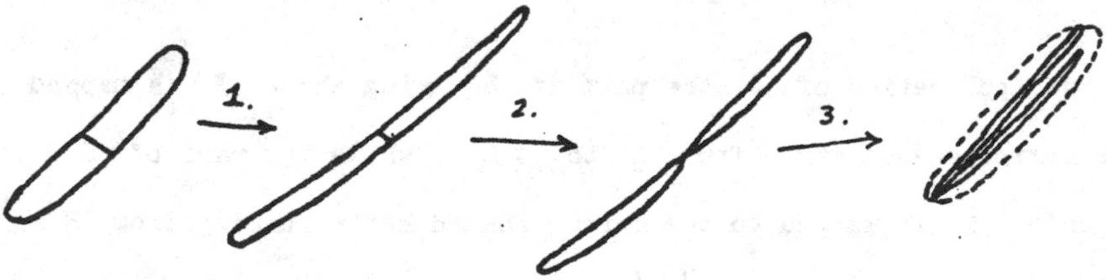

Figure 7

Straightening out and broadening the picture a bit, we obtain the shaded region below as the image of S under $\hat{\Phi}$

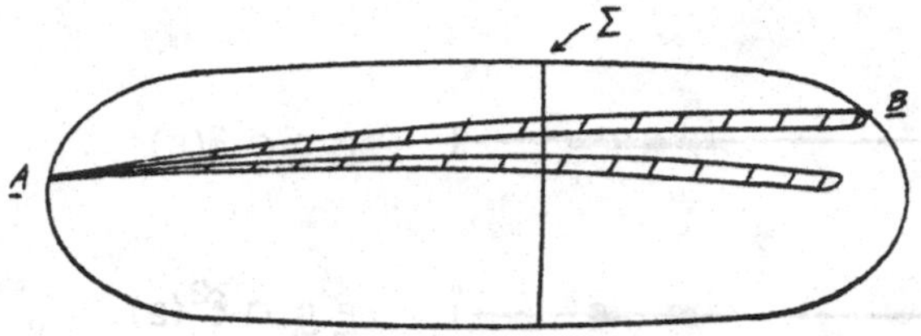

Figure 8

Here, Σ goes to $\underline{A}$ with pinching and $\underline{A}$ goes smoothly to $\underline{B}$. Applying $\hat{\Phi}$ again gives a set consisting of four long thin pieces, two inside the upper shaded region of Figure 8 and two inside the lower. All four strips are pinched together at $\underline{A}$ and the two upper ones are also pinched together at $\underline{B}$

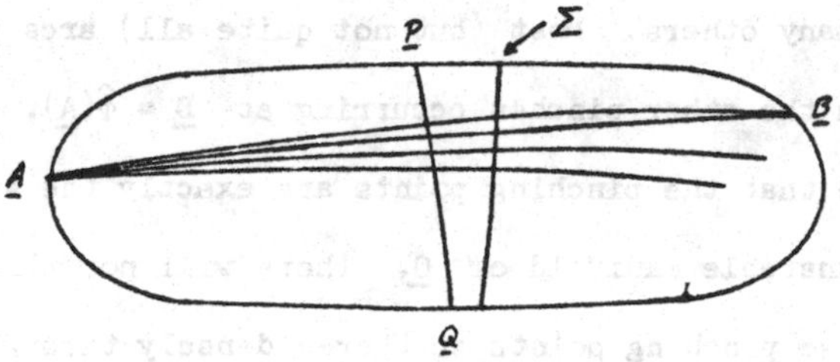

Figure 9

Similarly $\hat{\Phi}^3(S)$ consists of eight strips, all pinched together at $\underline{A}$, four pinched at $\underline{B}$, and two pinched at $\Phi(\underline{B})$. If we intersect with a transverse line $\underline{P}\,\underline{Q}$ we find successively

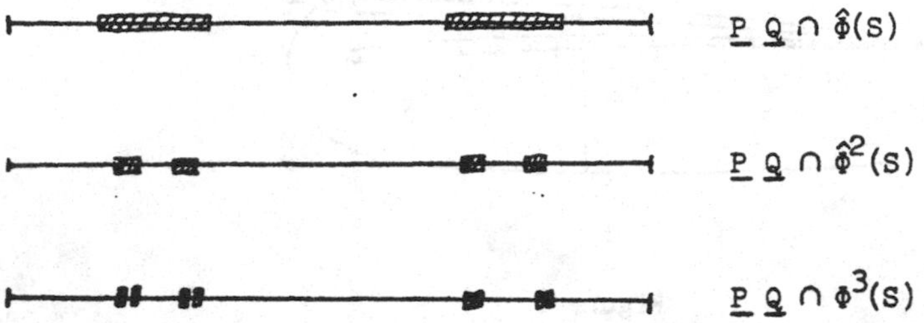

Figure 10

Continuing the process, $\hat{\Omega} = \bigcap_{n=1}^{\infty} \hat{\Phi}^n(S)$ consists of uncountably many longitudinal arcs and intersects any transverse arc like $\underline{P}\,\underline{Q}$ in a Cantor set. The longitudinal arcs making up $\hat{\Omega}$ are joined together in the following complicated way: Each arc is pinched together at each end with uncountably many others. Most (but not quite all) arcs pinch together at $\underline{A}$, with the other pinches occurring at $\underline{B} = \hat{\Phi}(\underline{A})$, $\hat{\Phi}^2(\underline{A}), \hat{\Phi}^3(A), \ldots$. Note that the pinching points are exactly the successive downcrossings of the unstable manifold of $\underline{0}$. There will normally be infinitely many of these pinching points scattered densely through $\hat{\Phi}$, although is it also possible for the set of pinching points to be finite. This latter situation happens if and only if $\underline{A}$ is a periodic point for $\hat{\Phi}$ or equivalently if the unstable manifold for $\underline{0}$ is contained in the stable manifold; we should expect this to be the case for a countable dense set of values of r in any small neighborhood of 28.

Whether the set of pinching points is finite or not, it is easy to see that $\hat{\Phi}$ has <u>some</u> orbit which is dense in $\hat{\Omega}$ and hence that $\hat{\Omega}$ is contained in the non-wandering set for $\hat{\Phi}$. Since $\hat{\Omega} = \bigcap_{n=1}^{\infty} \hat{\Phi}^n(S)$, $\hat{\Omega}$ attracts all orbits starting in S, so it meets all the requirements of our earlier provisional definition of attractor, (and it certainly deserves the epithet "strange.") The corresponding attractor for the solution flow T^t is now not hard to visualize. In the vicinity of the plane $Z = 27$ it consists locally of stacks of uncountably many two dimensional pieces which intersect transverse arcs in Cantor sets. Globally these two-dimensional leaves all pinch together along the unstable manifold of $\underline{0}$ consisting of two solution curves which we should expect to be dense in the attractor. Although the question has not been carefully investigated, it appears that the basin of attraction fills all of $\mathbb{R}^3$ except for $\underline{C}$, $\underline{C}'$ and their respective one-dimensional stable manifolds.

We now return, briefly, to the behavior of the Lorenz system for r slightly less than the critical value of 470/19. Because of the finite gap between the stationary solutions $\underline{C}$ and $\underline{C}'$ and the attractor, it is not really necessary for $\underline{C}$ and $\underline{C}'$ to be unstable in order for the attractor to exist; all that is necessary is that the unstable manifold of $\underline{0}$ which forms the outside edge of the attractor not fall into the basin of attraction of $\underline{C}$ or $\underline{C}'$. The existence of small unstable periodic orbits near $\underline{C}$ and $\underline{C}'$ shows that the basins of attraction of $\underline{C}$ and $\underline{C}'$ are not very large for r slightly below 470/19, and it turns out in fact that the unstable manifold of $\underline{0}$ is not attracted to

<u>C</u> and <u>C</u>′ unless r is less than about 24.1. Thus, for $24.1 < r < 24.74$, the system has (at least) three distinct attractors, the point attractors <u>C</u>, <u>C</u>′ and a strange attractor between them. Which attractor traps a given orbit depends on where the orbit starts, but orbits starting near <u>0</u> go to the strange attractor. Physically, the system displays <u>hysteresis</u>; it has several possible behaviors depending on its past history. If we imagine increasing the temperature gradient slowly from zero the solution will simply track one of the two stationary solutions up to $r = 470/19$. If, on the other hand, a temperature gradient making r slightly less than 470/19 is turned on suddenly with the system initially at rest, a state of permanent chaotic motion results.

In the above discussion, nothing has been said about the behavior of the Lorenz system for r larger than 28. Preliminary numerical experiments indicate that several further changes occur in the qualitative behavior of typical orbits, but, to my knowledge, a detailed analysis has not yet been made.

It may be interesting to note that the general structure of the Lorenz attractor — the fact that it is made from two-dimensional unstable manifolds of a pair of stationary solutions folded back on themselves infinitely often — does not depend on the dimensionality of the state space. A similar attractor can easily be constructed in a space of an arbitrary number of dimensions, and still consists locally of an uncountable family of two-dimensional sheets, stacked up in a Cantor-set-like way. It is thus at least possible that analogues of the Lorenz attractor exist for realistic approximations to the equations of hydrodynamics (or even for these equations themselves).

Chapter III. Ergodic Theory of Dissipative Systems

Let us now try to see what physical conclusions could be drawn if we knew that the full convection equations -- or some finite dimensional approximation to them which is sufficiently detailed to give an accurate description of the physical phenomena -- had behavior similar to that of the Lorenz system. Thus, consider a system of equations with an attractor on which the motion depends in a sensitive way on initial conditions and whose basin of attraction contains some physically relevant initial states. If the system is started out in the basin of attraction, its state at large positive times is not arbitrary: one can at least predict with confidence that it will be very near to the attractor, which will normally occupy a small fraction of the whole basin of attraction. On the other hand, because of the instability of the motion on the attractor itself, we cannot reasonably hope to be able to make accurate predictions about where near the attractor the system will be found. In other words, the state at large positive times is somewhat restricted as it must be near the attractor but otherwise appears to be "random," i.e., not to depend in a predictable way on the initial state.

As a practical matter, the main objective of the theory of convection is the computation of such quantities as the thermal conductivity of the convective layer, and these quantities are supposed to depend on the physical parameters of the system (viscosity, temperature gradient, etc.) but not on the initial state. At first glance it appears that these computations are impossible in principle if the asymptotic behavior is determined by something like the Lorenz attractor. The instantaneous rate of heat transfer can be expected to depend both on the time and on the initial state and is not likely to approach a limiting value at t goes to infinity. On closer examination, however, the situation is not as bad as it seems. What is usually required for applications is not, for example, the instantaneous rate of heat transfer, but rather the average of this quantity over a long period of time, and it is only the limiting value of this time average which needs to be independent of initial conditions. This suggests that it would be useful to have some sort of ergodic theorem for dissipative systems. We will now outline one possible version of such a theorem, motivated on the one hand by its intended applications and on the other hand by what has been proved in special cases.

Let T^t be a flow, Λ an attractor for T^t with basin of attraction B. By an ergodic theorem for (T^t,Λ) we mean a theorem asserting the existence of the following objects:

a) A probability measure μ_Λ on Λ, invariant under the solution flow T^t

b) A subset X of B, of Lebesgue measure zero such that:

For any continuous function f on B and any x in B but not in X,

$$\lim_{\tau\to\infty} \frac{1}{\tau}\int_0^\tau dt\ f(T^t x) = \int f\ d\mu_\Lambda .$$

This formulation has a number of related aspects; we offer the following remarks to clarify what it is intended to mean. The main thing being asserted is that forward time averages of "general" functions on the basin of attraction exist and are independent of the initial state. Independence of initial conditions cannot be expected to be true entirely without qualification. For example, most non-trivial attractors contain infinitely many unstable periodic orbits; the time average starting at a point exactly on one of these orbits will simply be the average over the orbit, which will not be at all like the time average for a typical initial point. We must therefore be prepared to throw out an exceptional set of initial conditions — in our formulation, the set X — which ought to be negligible from the physical point of view. We have taken as our criterion of physical negligibility that the set of exceptional points have Lebesgue measure zero. Note that Lebesgue measure itself has relatively little connection with the flow T^t, and in particular is not supposed to be invariant under T^t; it has rather been pulled in artificially to provide an elementary way of stating that a certain set is negligible. This criterion for negligibility has a number of drawbacks — notably, it applies only to flows on finite-dimensional manifolds and not to the convection equations themselves — and there are indications that it could be replaced by a sharper condition formulated in terms of Hausdorff dimension.

Next: We are considering time averages only for continuous functions and not, say, for general bounded Borel functions. Some such restriction is necessary to avoid trivial counterexamples arising from the fact that the flow is non-recurrent on $B \setminus \Lambda$. For example, if Λ consists simply of an attracting stationary solution, it is easy to construct a bounded Borel function f such that

$$\lim_{\tau\to\infty} \frac{1}{\tau}\int_0^\tau dt\ f(T^t x)$$

does not exist for any x in the basin of attraction other than the stationary solution itself.

Third: In our formulation, the exceptional set is taken to be independent of the function f, whereas it might seem more reasonable to allow it to vary with f. It turns out, however, to be no more restrictive to require the existence of a single exceptional set. To see this, assume that time averages exist and are independent of initial condition for each continuous function, but allow the exceptional set to depend on the function. Choose a countable set of continuous functions whose restrictions to Λ are dense in the space of all continuous functions on Λ; this will be possible, provided that Λ is compact. Let X be the union of the exceptional sets for these countably many functions; X will again be a set of Lebesgue measure zero and it is easy to see that time averages exist and are independent of initial condition in $B \setminus X$ for all continuous functions f.

Finally: Our formulation of a general ergodic theorem requires that time averages be obtained as mean values with respect to a probability measure μ_Λ on Λ. If Λ is compact, this is automatic once time averages are known to exist and to be essentially independent of initial condition. To see this, let $\bar{f}$ denote the common value of $\lim_{T\to\infty} \frac{1}{T}\int_0^T dt\, f(T^t x)$ for almost all x. The quantity $\bar{f}$ is defined for all functions f continuous on B, but is is easily seen that two functions which are equal on Λ have the same average, so $f \mapsto \bar{f}$ can be regarded as a functional defined on the space of continuous functions on Λ. This functional is linear, positive, and takes the constant function 1 to 1, and hence, by the Riesz Representation Theorem, has the form

$$f \mapsto \int f \, d\mu_\Lambda$$

for a uniquely determined probability measure μ_Λ on Λ. In spite of the fact that its existence is automatic, the measure μ_Λ is interesting and important since it ought to be possible to describe it intrinsically and hence to give a procedure for computing time averages other than by applying the definition. We have here a close analogy to the usual view of the role of the microcanonical ensemble in classical statistical mechanics, and the measure μ_Λ may therefore be viewed as an equilibrium ensemble for the dissipative system. One important practical difference from classical statistical mechanics should be noted: The microcanonical ensemble for a Hamiltonian system can be written down directly in terms of the Hamiltonian. To construct μ_Λ, on the other hand, it is necessary first to locate the attractor Λ and then to analyze exhaustively the behavior of the solution flow on and near Λ. So far, this process appears to require detailed information about the solutions to the equations of motion, as opposed to simply knowing the differential equations themselves.

To get a complete picture of the behavior of typical solutions of a set of differential equations, we would want to do something like the following:

a. show that, except for a set of Lebesgue measure zero, the state space splits into the basins of a finite number of attractors.

b. prove an ergodic theorem for each of these attractors.

The asymptotic properties of a solution curve will then depend on which basin of attraction it lies in, but essentially all solution curves in a given basin will have the same statistical properties over long periods of time. This program has been completely carried out by Ruelle and Bowen [10],[1] for flows on compact manifolds which satisfy Smale's Axiom A.

Rather than describe the proof of the Ruelle-Bowen theorem, we will try to illustrate the idea of the proof by showing how it might be adapted to prove an ergodic theorem for the Lorenz system. This procedure has the advantage of concreteness and relative simplicity; it has the disadvantage that it is not really a proof of anything as:

a. the argument starts from some qualitative features of the Lorenz attractor which are strongly suggested by numerical experiments but which are certainly not proved

b. even assuming these qualitative properties to hold, the proof of an ergodic theorem for the Lorenz system involves some algebraic and analytic complexities not present in the Axiom A case and not yet completely overcome.

What we will therefore actually do is to reduce the proof of the ergodic theorem of the Lorenz attractor to a question about a one-dimensional transformation and then suggest how to treat the one-dimensional problem by considering a model problem with a number of technical simplifications.

The first step in our proposed proof of an ergodic theorem for the Lorenz attractor is a reduction from the solution flow T^t to the Poincaré map Φ discussed in the preceding chapter. That is, we assume we have an ergodic theorem for Φ and show how to get one for T^t. For $\underline{x}$ in the plane $Z = 27$, let $\tau(\underline{x})$ denote the time required for the solution curve through $\underline{x}$ to return to its first downcrossing of the plane; if the solution curve never returns, we put $\tau(\underline{x}) = \infty$. For any $\underline{x}_1$ in the basin of attraction whose solution curve eventually makes a downcrossing at a point $\underline{x}$, time averages starting at $\underline{x}_1$ have the same limit as time averages starting at $\underline{x}$, so we may as well consider only time averages starting at points $\underline{x} = (X,Y,27)$ where $\frac{dZ}{dt} < 0$, i.e., where $X \cdot Y < 72$. We will exclude immediately $\underline{x}$'s on the stable manifold of $\underline{0}$; as the stable manifold is a set of measure zero, this will not affect the proof of an ergodic theorem. Then, for any continuous function f_1

$$\lim_{\tau\to\infty} \frac{1}{\tau}\int_0^{\tau} dt\ f_1(T^t\underline{x}) = \frac{\lim\limits_{N\to\infty} \frac{1}{N}\sum\limits_{n=0}^{N-1} f(\Phi^n\underline{x})}{\lim\limits_{N\to\infty} \frac{1}{N}\sum\limits_{n=0}^{N-1} \tau(\Phi^n\underline{x})}$$

(where $f(\underline{x}) = \int_0^{\tau(\underline{x})} f_1(T^t\underline{x})\ dt$)

provided both the limit in the numerator and limit in the denominator exist. If f and τ were continuous, an ergodic theorem for Φ would say that both numerator and denominator exist for almost all $\underline{x}$ and are essentially independent of $\underline{x}$; the same would then follow for the limit on the left. Unfortunately, $\tau(\underline{x})$ is not continuous; it approaches infinity as $\underline{x}$ approaches Σ. To complete the reduction properly therefore requires an approximation argument using some special properties of the equilibrium ensemble for Φ. This argument is inessential to the main outline of the proof; we will not give it.

Next, and purely to simplify the exposition, we will restrict consideration to those continuous functions f invariant under the symmetry $(X,Y,Z) \to (-X,-Y,Z)$. This permits us to consider only one of the two parts of the attractor for Φ, and to replace Φ by $\hat{\Phi}$. General continuous functions can be handled by a straightforward extension of the argument. For the remainder of this chapter we will always consider $\hat{\Phi}$ rather than Φ, and we will drop the $\hat{}$.

We must next examine in detail the action of Φ on and near the attractor Ω. The picture we want to develop is that some neighborhood of Ω decomposes into a one-parameter family of non-intersecting arcs running transverse to the attractor. The arcs are characterized by the property that each of them contracts to a point under repeated application of Φ, i.e., if $\underline{x}_1$, $\underline{x}_2$ are in the same arc then the distance from $\Phi^n\underline{x}_1$ to $\Phi^n\underline{x}_2$ goes to zero rapidly as n goes to $+\infty$. Accordingly, we will refer to them as <u>contracting</u> arcs.

The existence of contracting arcs is suggested by the fact that Φ compresses strongly in a direction transverse to the attractor. To see in more detail what is happening, let us look at a point $\underline{x}$ on or near the attractor, and a line segment $\overline{\alpha\beta}$ passing either above or below $\underline{x}$ and roughly parallel to the attractor:

Figure 1

Applying Φ moves both $\underline{x}$ and $\overline{\alpha\beta}$ much closer to the attractor and also slides α and β slightly away from $\underline{x}$ along the attractor. Since α and β move away from $\underline{x}$ in opposite directions, there must be points like γ on $\overline{\alpha\beta}$ such that the separation between $\Phi\underline{x}$ and $\Phi\gamma$ remains at a substantial angle to the attractor. Because of the strength of the transverse compression of Φ, this condition locates γ fairly precisely along $\overline{\alpha\beta}$. Now apply Φ again and require that the separation between $\Phi^2\underline{x}$ and $\Phi^2\gamma$ remain transverse to the attractor; this will locate γ even more precisely. Continuing in this way we construct a sequence of successive approximations which ultimately yields a single point γ on $\overline{\alpha\beta}$ with the property that the separation between $\Phi^n\underline{x}$ and $\Phi^n\gamma$ is transverse to the attractor for all positive n. Because of this transversal separation and the fact that Φ compresses in the transverse direction, the distance from $\Phi^{n+1}\underline{x}$ to $\Phi^{n+1}\gamma$ is for each n a small fraction of the distance from $\Phi^n\underline{x}$ to $\Phi^n\gamma$, so the distance from $\Phi^n\gamma$ to $\Phi^n\underline{x}$ decreases exponentially with n. Any point of $\overline{\alpha\beta}$ other than γ, on the other hand, will eventually be drawn away from $\underline{x}$ by the stretching action of Φ along the attractor. For a fixed $\underline{x}$, the points γ on the various possible nearby longitudinal segments $\overline{\alpha\beta}$ string together to form a one-dimensional set which, by construction, is contracted under the action of Φ. Thus: Each point $\underline{x}$ sufficiently near to Ω ought to lie on a contracting arc. Contracting arcs are uniquely determined locally, and two contracting arcs which intersect must be continuations of each other. There is no apparent reason why contracting arcs must be unreasonably short; it ought to be possible to continue each of them at least across the full thickness of the attractor. We thus arrive at a picture like the following, where the predominantly vertical segments represent contracting arcs.

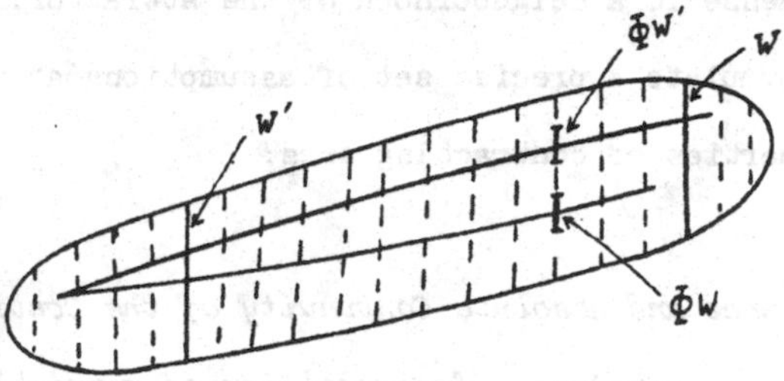

Figure 2

This figure illustrates an important feature of the decomposition into contracting arcs. If W is a contracting arc, cut off at the ends as illustrated, then ΦW will be part but not all of another contracting arc. Frequently, there will be a second contracting arc W', running across the opposite end of the attractor, such that ΦW' shares a contracting arc with ΦW. In this case, if $\underline{x} \in W$

and $\underline{x}' \in W'$ then the distance from $\Phi^n \underline{x}$ and $\Phi^n \underline{x}'$ goes to zero exponentially as n goes to infinity in spite of the fact that $\underline{x}, \underline{x}'$ are not in the same contracting arc. For $\underline{x}'$ to be in the same contracting arc as $\underline{x}$ it is necessary but <u>not</u> sufficient to have $\Phi^n \underline{x}$ and $\Phi^n \underline{x}'$ approach each other as n goes to infinity. It is in fact to be expected that, for a typical point $\underline{x}$ near the attractor, the set of points $\underline{x}'$ near the attractor such that

$$\lim_{n\to\infty} d(\Phi^n \underline{x}, \Phi^n \underline{x}') = 0$$

will consist of an infinite (but countable) union of contracting arcs and will be dense in a neighborhood of the attractor.

We can now formulate a precise set of assumptions about the existence and properties of contracting arcs:

Assumption: Existence and Absolute Continuity of the Contracting Foliation. There is a continuous decomposition of a neighborhood of the attractor Ω *into a one-parameter family of smooth arcs (contracting arcs) with the following properties:*

a. (Contractivity) There exist constants C, λ, *with* $0 < \lambda < 1$, *such that if* x_1, x_2 *are in the same contracting arc*

$$d(\Phi^n \underline{x}_1, \Phi^n \underline{x}_2) < C\lambda^n \qquad n = 0,1,2,\ldots$$

b. (Invariance) The image under Φ of a contracting arc is contained in a contracting arc.

It would now be natural to construct a co-ordinate system for a neighborhood of Ω such that the contracting arcs are lines where one of the co-ordinates is constant. It turns out ultimately to be more convenient to do only part of the reparametrization, i.e. to construct only the co-ordinate which is constant on contracting arcs or, equivalently, to parametrize the set of contracting arcs. To do this we draw in some convenient way a smooth arc γ running the full length of the attractor (but not necessarily in the attractor) which crosses each contracting arc exactly once and at a non-zero angle. (We will refer to such an arc γ as a <u>longitudinal arc</u>). We let π denote the projection of a neighborhood of the attractor onto γ along contracting arcs, i.e., for each $\underline{x}$, $\pi(\underline{x})$ denotes the unique point of γ on the same contracting arc as $\underline{x}$.

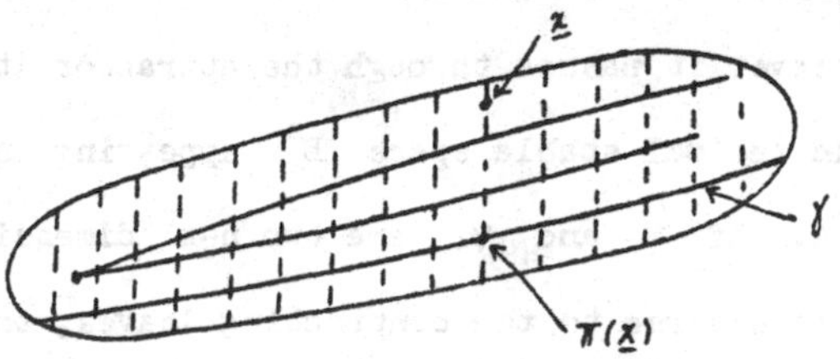

Figure 3

We add to our list of assumptions a regularity property for π:

c. (Absolute continuity) If γ_1 *is any other longitudinal arc, then* π *restricted to* γ_1 *is a differentiable mapping from* γ_1 *to* γ *with a Hölder continuous derivative.*

This does not complete the statement of the assumption, but at this point we digress to relate these conditions to known facts about Axiom A attractors. (It perhaps needs to be mentioned that Φ does not satisfy Axiom A; it is neither one-one nor differentiable on Σ. Its apparently mild failure to fulfill the conditions turns out to have far-reaching consequences; the Lorenz attractor has a much more intricate and delicate structure than is possible for an Axiom A attractor.) For a general Axiom A attractor, each point of the attractor has a neighborhood (in the manifold of states) which splits continuously into an $(n-n^s)$-parameter family of smooth submanifolds of dimension n^s, called contracting leaves and analogous to the contracting arcs of the above discussion. Each contracting leaf shrinks exponentially under repeated applications of the transformation; moreover, wherever it passes through the attractor it is tangent to the infinitesimal stable space E^s appearing in the statement of Axiom A. If γ and γ_1 are two $n-n^s$ dimensional surfaces each running transverse to the contracting leaves, then projection along contracting leaves defines a continuous mapping from γ_1 to γ. One of the unpleasant technical features of this subject is the fact that, even if the transformation itself is infinitely differentiable, this projection does not need to be continuously differentiable. It is, however, sufficiently well behaved to send $(n-n^s)$-dimensional Lebesgue measure on γ_1 to the product of Lebesgue measure on γ with a Hölder continuous density. This property is called absolute continuity of the contracting foliation; in the special case where $n^s = n-1$ (so γ, γ_1 are one-dimensional), absolute continuity implies continuous differentiability.

We return now to the problem of proving an ergodic theorem for Φ. The next step in the argument is to show that it suffices to prove the existence and essential independence of initial point for functions f which are continuous and constant on contracting arcs. To see this, we assume the ergodic theorem for such functions and prove it for general continuous functions. Thus, let f be uniformly continuous on a neighborhood V of the attractor which splits, as in the above assumption, into contracting arcs. Choose $\varepsilon > 0$, and find $\delta > 0$ such that $d(x_1,x_2) \leq \delta$ implies $|f(x_1) - f(x_2)| \leq \varepsilon/2$. Then choose m large enough so that the image under Φ^m of any contracting arc has diameter smaller than δ. The variation of $f\circ\Phi^m$ over any contracting arc is thus no larger than $\varepsilon/2$, and from this it follows easily that there exists a continuous function g, constant on contracting arcs, which differs from $f\circ\Phi^m$ by no more than ε everywhere on V. By assumption

$$\lim_{N\to\infty} \frac{1}{N} \sum_{n=0}^{N-1} g(\Phi^n \underline{x})$$

exists and is equal to a constant almost everywhere; call the constant C_g. Hence

$$\limsup_{N\to\infty} \frac{1}{N}\sum_{n=0}^{N-1} f\circ\Phi^{n+m} \quad \text{and} \quad \liminf_{N\to\infty} \frac{1}{N}\sum_{n=0}^{N-1} f\circ\Phi^{n+m}$$

both differ from C_g by less than ε almost everywhere. Changing summation indices and making some simple estimates shows that

$$\limsup_{N\to\infty} \frac{1}{N}\sum_{n=0}^{N-1} f\circ\Phi^{n} \quad \text{and} \quad \liminf_{N\to\infty} \frac{1}{N}\sum_{n=0}^{N-1} f\circ\Phi^{n}$$

also differ from C_g by less than ε almost everywhere. Letting ε approach zero, we see that

$$\lim_{N\to\infty} \frac{1}{N}\sum_{n=0}^{N-1} f\circ\Phi^{n}$$

exists and is constant almost everywhere on V as desired.

We need now consider only functions f on V which are continuous and constant on contracting arcs, and the next step is to exploit the projection π along contracting arcs to reduce to a one-dimensional problem. Choose a smooth longitudinal arc as in part c. of our assumption above, and let $\tilde{f}$ denote the restriction of f to γ. Constancy of f along contracting arcs means that

$$f = \tilde{f}\circ\pi \quad .$$

Define a mapping $\tilde{\Phi}$ of γ to itself by

$$\tilde{\Phi} = \pi \circ \Phi|_{\gamma}$$

(i.e., first apply Φ, then project back to γ along contracting arcs).

By the invariance of the decomposition into contracting arcs

$$f \circ \Phi^n = (\tilde{f} \circ \tilde{\Phi}^n) \circ \pi \qquad n = 0,1,2,\ldots$$

i.e., the action of Φ on f is obtained by "lifting" the action of $\tilde{\Phi}$ on $\tilde{f}$. We now claim that, in order for

$$\lim_{N\to\infty} \frac{1}{N} \sum_{n=0}^{N-1} f \circ \Phi^n$$

to exist and be constant almost everywhere on V, it suffices that

$$\lim_{N\to\infty} \frac{1}{N} \sum_{n=0}^{N-1} \tilde{f} \circ \tilde{\Phi}^n$$

exist and be constant almost everywhere on γ with respect to one-dimensional Lebesgue measure. Indeed, if

$$\lim_{N\to\infty} \frac{1}{N} \sum_{n=0}^{N-1} \tilde{f} \circ \tilde{\Phi}^n = C_f$$

everywhere except on a set $\tilde{X}$ of linear measure zero, then

$$\lim_{N\to\infty} \frac{1}{N} \sum_{n=0}^{N-1} f\circ\Phi^n = C_f$$

everywhere on V except on $\pi^{-1}\tilde{X}$. By the absolute continuity of the decomposition into contracting arcs, $\pi^{-1}\tilde{X}$ intersects any longitudinal arc in a set of linear measure zero. From this and Fubini's Theorem it follows readily that $\pi^{-1}\tilde{X}$ is a set of two-dimensional measure zero.

We have thus reduced the proof of an ergodic theorem for Φ to the proof of a corresponding statement for the mapping $\tilde{\Phi}$ of γ to itself. By introducing a smooth parametrization for γ we may transport everything to the unit interval. For definiteness we take the parametrization so that zero corresponds to the end of γ near the stationary solution $\underline{C}$, and we will denote by φ the mapping of the unit interval to itself obtained from $\tilde{\Phi}$ by introducing coordinates in this way. Since φ is a mapping of the unit interval to itself, it can be represented by a graph, and its graph looks schematically like

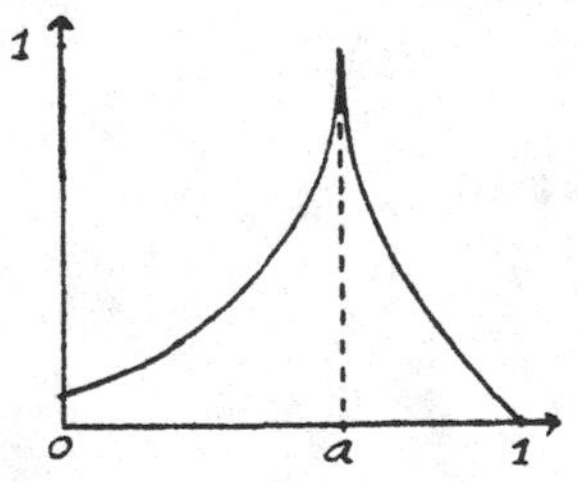

Figure 4

where a corresponds to the point on γ where it crosses Σ. There is a cusp at a; otherwise, φ is continuously differentiable with a Hölder-continuous first derivative. We cannot expect φ to be infinitely differentiable away from a, since it is obtained by introducing coordinates for $\tilde{\Phi} = \pi \circ \Phi$ and, although Φ is infinitely differentiable away from Σ, π cannot be expected to be very smooth. We now add a final assumption which makes precise the intuitive notion that Φ stretches in the longitudinal direction.

d. (Longitudinal expansiveness). The parametrization for γ may be chosen in such a way that φ' is strictly greater than one on $[0,a)$ *and strictly less than minus one on* $(a,1]$.

Given this assumption, and the qualitative properties of described above, we would like to prove that, for any continuous function f on $[0,1]$,

$$\lim_{N\to\infty} \frac{1}{N} \sum_{n=0}^{N-1} f \circ \varphi^n$$

exists and is constant almost everywhere with respect to Lebesgue measure on $[0,1]$. The next step in our chain of reductions is to reduce this statement to one about invariant* measures for φ. To

---

*Recall that a measure μ is said to be invariant under a measurable mapping φ (which need not be invertible) if $\mu(\varphi^{-1}E) = \mu(E)$ for all measurable sets E.

see how the reduction goes, let us imagine for a moment that Lebesgue measure was invariant under φ. Then the existence of

$$\lim_{N\to\infty} \frac{1}{N} \sum_{n=0}^{N-1} f\circ\varphi^n$$

almost everywhere would follow immediately from Birkhoff's Pointwise Ergodic Theorem; if Lebesgue measure were furthermore ergodic, then constancy of this limit almost everywhere would also be automatic. Although Lebesgue measure is not invariant under φ, the same argument could be applied if we knew that there existed some invariant measure equivalent to (i.e., with the same sets of measure zero as) Lebesgue measure. Thus: Granting the assumptions made to justify the above reductions, proving an ergodic theorem for the Lorenz attractor reduces to proving:

If φ *is a mapping of* $[0,1]$ *to itself with the qualitative features described above, then there is a probability measure* μ *equivalent to Lebesgue measure on* $[0,1]$ *which is invariant and ergodic with respect to* φ.

Proof of this statement is complicated by some unavoidable but secondary features of the mapping φ. We will therefore present the proof for a simpler class of mappings which display in a more transparent way the techniques used for investigating existence and uniqueness of absolutely continuous invariant measure and the relation between such measures and Gibbs states for one-dimensional lattice systems. The mappings we will consider are again continuous mappings φ of the unit interval to itself with the following properties:

a) For some $a \in (0,1)$, φ is monotonic increasing on $(0,a)$, decreasing on $(a,1)$.

b) $\varphi(a) = 1, \varphi(1) = 0$

c) φ is differentiable on $[0,a)$ and on $(a,1]$; its derivative is Hölder continuous on each of these intervals and approaches a finite limit when a is approached either from the left or from the right.

d) For some $\alpha < 1$

$$|\varphi'(x)| > \alpha^{-1} \qquad \text{for all } x \text{ (except } a)$$

e) $\varphi(0) = a$.

Thus, the graph of φ looks like:

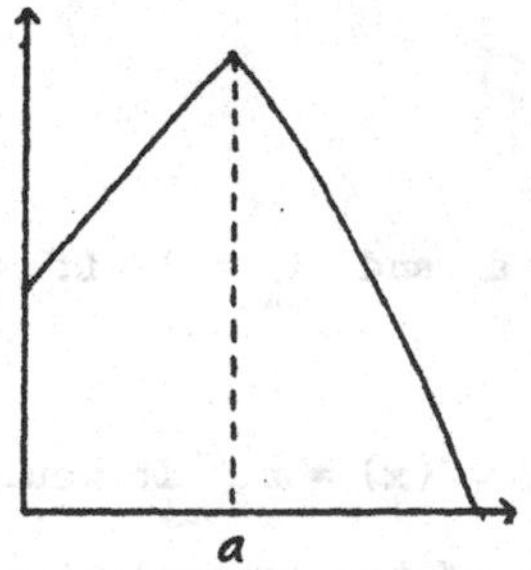

Figure 5

Our model φ differs from the φ derived from the Lorenz model in two respects. The more important is the absence of a cusp at a; the second is the requirement e) that $\varphi(0) = a$. At the expense of only slight complications, the latter condition could be replaced by the more realistic requirement that, for some i_0, we have

$$0 < \varphi(0) < \varphi^2(0) < \ldots < \varphi^{i_0}(0) = a .$$

For the Lorenz system, on the other hand, there is no reason to believe that any $\varphi^i(0)$ is exactly equal to a, and this fact introduces an extra layer of complexity.

From now on, then, we consider a mapping φ satisfying a) - e). As an aid to analyzing such a mapping, we introduce a coding of [0,1] into a set of (one-sided infinite) sequences of zeros and ones. Intuitively, we want to associate with each x a sequence $(i_0, i_1, i_2, \ldots)$ where

$$i_k = 0 \text{ if } \varphi^k(x) \geq a \text{ and } i_k = 1 \text{ if } \varphi^k(x) \leq a.$$

The prescription is ambiguous if $\varphi^k(x) = a$. It would be easy to lift this ambiguity by making one of the inequalities strict, but for our purposes it is better to allow such x's to have more than one coding. The inverse correspondence - from sequences to x's -

will turn out to be unambiguously defined and almost, but not quite, one-one. We set

$$\Delta(0) = [a,1],\ \Delta(1) = [0,a],$$

so $(i_0,\dots)$ is a coding of x if $\varphi^k(x) \in \Delta(i_k)$ for $k = 0,1,2,\dots$. It follows readily from the assumed properties of φ that 0 is the only element x of $\Delta(1)$ with $\varphi(x) \in \Delta(1)$ and, since $\varphi^3(0) = 1$, that there is no x with $x \in \Delta(1)$, $\varphi(x) \in \Delta(1)$, $\varphi^2(x) \in \Delta(1)$. Thus, no coding sequence has three successive ones and only x's such that $\varphi^k(x) = 0$ for some k admit codings with two successive ones. In the latter case, replacing the block (1,1) wherever it appears in the coding sequence of the block (1,0) gives another coding of the same point, so:

Every point of [0,1] *admits a coding without successive ones.*

We will say that a finite or infinite sequence of 0's and 1's is <u>admissible</u> if it contains no pair of successive 1's, and we will let $\mathcal{A}$ denote the set of all admissible sequences. *In what follows we consider only codings into admissible sequences.*

The next step in the analysis is to show that every admissible sequence is a coding of exactly one x. To do this, we define, for any finite admissible sequence $(i_0,i_1,\dots,i_k)$,

$$\Delta(i_0,i_1,\ldots,i_k) = \{x:\varphi^j(x) \in \Delta(i_j) \quad \text{for} \quad j = 0,1,\ldots,k\}$$

(i.e., $\Delta(i_0,i_1,\ldots,i_k)$ is the set of x's admitting codings which begin with $(i_0,\ldots,i_k)$.)

Proposition. For any finite admissible sequence $(i_0,i_1,\ldots,i_k)$, $\Delta(i_0,\ldots,i_k)$ *is a non-empty closed interval of length no larger than* α^k *(where* α^{-1} *is the expansion constant of condition d) above).*

Proof. We argue by induction on k. The statement is clearly true for $k = 0$. For $k > 0$ we have by definition:

$$\Delta(i_0,\ldots,i_k) = \Delta(i_0) \cap \varphi^{-1}(\Delta(i_1,\ldots,i_k))$$

From the assumed properties of φ it follows that φ is one-one, continuously differentiable, and expanding on $\Delta(i_0)$, and that $\varphi(\Delta(1)) = \Delta(0)$; $\varphi(\Delta(0)) = [0,1]$. Thus, since $(i_0,i_1) \neq (1,1)$, $\varphi(\Delta_{i_0}) \supset \Delta(i_1) \supset \Delta(i_1,\ldots,i_k)$ so φ maps $\Delta(i_0,\ldots,i_k)$ onto $\Delta(i_1,\ldots,i_k)$. By the induction hypothesis $\Delta(i_1,\ldots,i_k)$ is a non-empty closed interval, so the same is true of $\Delta(i_0,\ldots,i_k)$. Since $|\varphi'| > \alpha^{-1}$ on $\Delta(i_0)$,

$$\alpha^{-1}\lambda[\Delta(i_0,\ldots,i_k)] \leq \lambda[\Delta(i_1,\ldots,i_k)]$$

where λ denotes length of an interval (or Lebesgue measure of a more general set). Again using the induction hypothesis

$$\lambda[\Delta(i_1,\ldots,i_k)] < \alpha^{k-1},$$

so

$$\lambda[\Delta(i_0,\ldots,i_k)] < \alpha^{k}$$

as desired. ∎

If, now, $\underline{i} = (i_0,i_1,i_2,\ldots)$ is an admissible sequence,

$$\Delta(i_0) \supset \Delta(i_0,i_1) \supset \Delta(i_0,i_1,i_2) \supset \ldots$$

is a decreasing sequence of closed intervals with length going to zero and hence there is exactly one point x in every $\Delta(i_0,i_1,\ldots,i_k)$. We will note this x by $\pi(\underline{i})$; it evidently admits $\underline{i}$ as a coding and is the only point which does. Thus, every admissible sequence $\underline{i}$ is a coding of a uniquely defined point $\pi(\underline{i})$ in $[0,1]$. The mapping π is continuous from $\mathcal{A}$, equipped with the topology it inherits as a subset of the compact product space $\{0,1\}^{\mathbb{N}}$, to $[0,1]$. Although π is not one-one, it is easy to see that it is at most two-to-one and that there are only countably many x's with more than one admissible coding. We let σ denote the shift mapping on $\mathcal{A}$,

$$\sigma((i_0,i_1,\ldots)) = (i_1,\ldots);$$

then if $\underline{i}$ is an admissible coding of x, $\sigma\underline{i}$ is an admissible coding of $\varphi(x)$, i.e.

$$\pi(\sigma(\underline{i})) = \varphi(\pi(\underline{i}))$$

If π were exactly one-one, it would set up an isomorphism between φ and the shift mapping σ. Although this is not the case, π is close enough to a true isomorphism for many purposes, and this is true in particular for the analysis of continuous* measures on [0,1]. Any probability measure $\tilde{\mu}$ on $\mathcal{A}$ projects under π to a probability measure $\pi\tilde{\mu}$ on [0,1] defined by

$$(\pi\tilde{\mu})(E) = \tilde{\mu}(\pi^{-1}E).$$

The inverse operation -- lifting measures from [0,1] to $\mathcal{A}$-- is not quite so simple, but to any continuous probability measure μ on [0,1] there corresponds a unique continuous probability measure

---

*A measure is <u>continuous</u> if it assigns measure zero to any set consisting of a single point.

$\tilde{\mu}$ on $\mathcal{A}$ with $\pi\tilde{\mu} = \mu$. To construct $\tilde{\mu}$, we let $\tilde{\Delta}(i_0,\ldots,i_k)$ denote the cylinder set

$$\{\underline{j}\colon\ j_0 = i_0,\ldots,j_k = i_k\};$$

we define $\tilde{\mu}$ on cylinder sets by

$$\tilde{\mu}(\tilde{\Delta}(i_0,\ldots,i_k)) = \mu(\Delta(i_0,\ldots,i_k));$$

and we use standard measure theory to extend $\tilde{\mu}$ to a Borel probability measure on $\mathcal{A}$. (The construction does not work for a completely general measure μ because $\tilde{\mu}$ as defined above will not be finitely additive if the boundaries of the $\Delta(i_0,\ldots,i_k)$ have non-zero μ-measure. For continuous measures there is no problem since the boundary of $\Delta(i_0,\ldots,i_k)$ contains only two points.) Projection by π, then, sets up an isomorphism between the sets of continuous probability measures on $\mathcal{A}$ and on $[0,1]$. This isomorphism is easily seen to preserve most interesting relations, e.g. $\tilde{\mu}_1$ and $\tilde{\mu}_2$ are equivalent if and only if $\pi\tilde{\mu}_1$ and $\pi\tilde{\mu}_2$ are, and $\tilde{\mu}$ is invariant (ergodic) under σ if and only if $\pi\tilde{\mu}$ is invariant (ergodic) under φ.

We can therefore adopt the following strategy for constructing a φ-invariant measure equivalent to Lebesgue measure λ:

i) Lift Lebesgue measure to a measure $\tilde{\lambda}$ on $\mathcal{A}$

ii) Construct a shift invariant measure $\tilde{\mu}$ on $\mathcal{A}$ which is ergodic and equivalent to $\tilde{\lambda}$

iii) Project $\tilde{\mu}$ under π to obtain the desired measure μ on $[0,1]$.

This strategy involves a trade-off. It replaces the possibly complicated mapping φ by the simple and standard shift mapping σ, but it also replaces the simple and standard Lebesgue measure by the less simple measure $\tilde{\lambda}$ on $\mathcal{A}$. The utility of the trade-off depends on whether or not we can find methods to control the behavior of $\tilde{\lambda}$. The key to obtaining such control is the observation that $\tilde{\lambda}$ is the Gibbs state for a one-dimensional semi-infinite classical lattice system with a rapidly-decreasing shift invariant many-body potential. (By semi-infinite we mean that the lattice sites are labelled by the non-negative integers rather than by all the integers.) Since the configuration space is $\mathcal{A}$ — the set of sequences of 0's and 1's with no two consecutive 1's — the lattice system will have a nearest neighbor exclusion, but we will argue shortly that the potential is otherwise finite.

Before justifying the claim that $\tilde{\lambda}$ is a Gibbs state, let us first describe why this fact is useful. Standard theorems about one-dimensional classical statistical mechanics can be applied to show that the same interaction for the two-sided infinite lattice system has a unique Gibbs state and that this Gibbs state is translation (i.e., shift) invariant with very good ergodic properties. (See Ruelle [9] for uniqueness, Gallavotti [2] for ergodic properties.) Let $\tilde{\mu}$ be the measure on $\mathcal{A}$ obtained from the invariant Gibbs state by ignoring the part of the lattice system to the left of the origin (i.e., by projection). Using the fact that the interaction between the part of the lattice system to the left of the origin and the part to the right is bounded except for the effects of the nearest neighbor exclusion, it is easy to show that $\tilde{\mu}$ is absolutely continuous with respect to $\tilde{\lambda}$ with a Radon-Nikodym derivative which is both bounded above and bounded away from zero. Thus, once $\tilde{\lambda}$ has been identified as a Gibbs state, the standard theory of Gibbs states yields almost immediately the existence and ergodicity of a shift-invariant measure $\tilde{\mu}$ equivalent to $\tilde{\lambda}$ and hence of a φ-invariant measure equivalent to Lebesgue measure.

To see why $\tilde{\lambda}$ is a Gibbs state, let us fix an admissible sequence $i_1, i_2, \ldots$ and compute the conditional probabilities with respect to $\tilde{\lambda}$ of the two possible values - 0 and 1 - of i_0 given $i_1, i_2, \ldots$. (To complete the identification, we will need to compute, more generally, conditional probabilities of the various values of $i_0, \ldots, i_k$ given $i_{k+1}, \ldots$.) If $i_1 = 1$ the computation is trivial; i_0 has to be 0 since the sequence $i_0, i_1, \ldots$ would otherwise not be allowed. We assume therefore that $i_1 = 0$. In this case the conditional probability is equal to

$$\lim_{m\to\infty} \frac{\tilde{\lambda}(\tilde{\Delta}(i_0,\ldots,i_m))}{\tilde{\lambda}(\tilde{\Delta}(0,i_1,\ldots,i_m)) + \tilde{\lambda}(\tilde{\Delta}(1,i_1,\ldots,i_m))}$$

(More precisely: General theory assures us that this limit exists for almost all $(i_1,i_2,\ldots)$ and is equal almost everywhere to the desired conditional probability. We will in fact show that the limit exists for all $(i_1,i_2,\ldots)$ and give a formula for it.)

Because $\tilde{\lambda}$ is obtained by transporting Lebesgue measure,

$$\tilde{\lambda}(\tilde{\Delta}(i_0,\ldots,i_m)) = \lambda(\Delta(i_0,\ldots,i_m)).$$

Recall that $\Delta(i_0,\ldots,i_m)$ is an interval of length no greater than α^m. Moreover, for each of the two possible values of i_0, φ maps $\Delta(i_0,\ldots,i_m)$ onto $\Delta(i_1,\ldots,i_m)$. If m is large, and if we write x_{i_0} for $\pi(i_0,i_1,\ldots)$ then φ' is nearly equal to $\varphi'(x_{i_0})$ on all of $\Delta(i_0,i_1,\ldots,i_m)$ and we have

$$\lambda(\Delta(i_1,\ldots,i_m)) \simeq |\varphi'(x_{i_0})|\,\lambda(\Delta(i_0,\ldots,i_m)).$$

The approximation becomes exact as $m \to \infty$ so

$$\lim_{m\to\infty} \frac{\lambda(\Delta(0,i_1,i_2,\ldots))}{\lambda(\Delta(1,i_1,i_2,\ldots))} = \frac{|\varphi'(\pi(1,i_1,i_2,\ldots))|}{|\varphi'(\pi(0,i_1,\ldots\quad))|}$$

Thus if we define

$$h(i_0,i_1,i_2,\ldots) = \log |\varphi'(\pi(i_0,i_1\ldots))|$$

we find that the conditional probability of i_0 given $i_1,i_2,\ldots$ is equal to

$$e^{-h(i_0,i_1,\ldots)}/\{e^{-h(0,i_1,\ldots)} + e^{-h(1,i_1,\ldots)}\}$$

Entirely similar arguments show that the conditional probability of $i_0,\ldots,i_k$ given $i_{k+1},\ldots$ is equal to

$$\frac{\exp[-h(i_0,i_1,\ldots) - h(i_1,i_2,\ldots) - \ldots - h(i_k,i_{k+1},\ldots)]}{\sum_{i'_0,\ldots,i'_k} \exp[-h(i'_0,\ldots,i'_k,i_{k+1},\ldots) - \ldots - h(i'_k,i_{k+1},\ldots)]}$$

We can now construct the interaction for which $\tilde{\lambda}$ is a Gibbs state. The intuitive idea is that $h(i_0,i_1,\ldots)$ should represent the contribution of the lattice site zero to the total energy. This is not a well-defined concept, however, so there will be some choices to be made in the construction of the interaction. We will think of our lattice system as a spin system rather than a lattice gas. The interaction is then specified by giving, for each finite subset X of $\mathbf{Z}$, a function Φ_X defined on $\{0,1\}^X$, subject to a straightforward translation-invariance requirements; Φ_X is interpreted as the potential energy due to direct interaction among all the lattice sites in X. The total energy for a configuration $\underline{i}$ defined on a finite sublattice Λ is

$$U_\Lambda(\underline{i}) = \sum_{X\subset\Lambda} \Phi_X(\underline{i}|_X)$$

We construct an interaction by defining:

$\Phi_X = 0$ unless X is an interval (i.e., a set of the form $(j, j+1, \ldots, j+k)$

$$\Phi_{\{0,1,\ldots,k\}}(i_0,\ldots,i_k) = H_k(i_0,\ldots,i_k) - H_{k-1}(i_0,\ldots,i_{k-1})$$

where

$$H_k(i_0,\ldots,i_k) = \inf_{i'_{k+1},\, i'_{k+2},\ldots} h(i_0,\ldots,i_k,i'_{k+1},i'_{k+2},\ldots), k = 0,1,\ldots$$

$$H_{-1} = 0$$

The function Φ_X for an interval of length $k+1$ not starting at zero is determined by translation invariance:

$$\Phi_{\{j,j+1,\ldots,j+k\}}(i_j,i_{j+1},\ldots,i_{j+k}) = \Phi_{\{0,1,\ldots,k\}}(i_j,\ldots,i_{j+k})$$

With these definitions it is easy to see that

$$U_{\{0,\dots,m\}}(i_0,\dots,i_m) = H_m(i_0,\dots,i_m) + H_{m-1}(i_1,\dots,i_m) + \dots + H_0(i_m)$$

$$U_{\{0,\dots,k\}}(i_0,\dots,i_k) + W(i_0,\dots,i_k | i_{k+1},\dots,i_m)$$

$$= H_m((i_0,\dots,i_m) + H_{m-1}(i_1,\dots,i_m) = \dots$$

$$+ H_{m-k}(i_k,i_{k+1},\dots,i_m).$$

Taking the limit of the second of these equations as $m \to \infty$ with k held fixed gives

$$U_{\{0,\dots,k\}}(i_0,\dots,i_k) + W(i_0,\dots,i_k | i_{k+1},\dots)$$

$$= h(i_1,i_2,\dots) + \dots + h(i_k,i_{k+1},\dots)$$

Comparing this equation with the previously obtained formula for conditional probabilities relative to $\tilde{\lambda}$ shows that $\tilde{\lambda}$ is indeed a Gibbs state for a semi-infinite lattice system with the interaction we have constructed.

To apply standard results from statistical mechanics to show that this interaction has a unique Gibbs state, we need to know that the interaction drops off rapidly at large separations. It turns out that

$$\max_{i_0,\dots i_k} |\Phi_{\{0,\dots,k\}}(i_0,\dots,i_k)|$$

goes to zero exponentially as k goes to infinity. To see this, observe:

a. $|\Phi_{\{0,\dots,k\}}(i_0,\dots,i_k)|$ is no larger than the maximum minus minimum of $\log|\varphi'(x)|$ on $\Delta(i_0,\dots,i_k)$

b. The length of $\Delta(i_0,\dots,i_k)$ is no larger than α^k

c. φ' is Hölder continuous on $\Delta(0)$ and on $\Delta(1)$.

To conclude, let us survey briefly how the above development would have to be modified to apply to the φ which comes from the Lorenz system. The Lorenz φ is again increasing to some point a; then decreasing from a to 1, and we have

$$\varphi(0) > 0; \ \varphi(a) = 1; \ \varphi(1) = 0.$$

Formally, we can approach this mapping in the same way as our model φ: We code each point x of $[0,1]$ into a sequence of ones and zeros determined by whether the successive $\varphi^k(x)$'s are to the left or the right of a, and thus translate the problem into a statistical-mechanical one. The technical complications are two-fold. First of all, the image of the coding is no longer as simple as before. It is possible to have more than two successive ones, but because φ moves points of $[0,a]$ non-trivially to the right it is not possible to have arbitrarily many. (The maximum number is actually 25 for $r = 28$). A straightforward analysis shows that, unless $\varphi^k(0) = a$ for some k, the image of the coding cannot be described by specifying a finite number of excluded finite sequences. In statistical mechanical terms, this means that the corresponding classical lattice system has infinitely many "exclusions," of arbitrarily long range, generalizing the nearest-neighbor exclusion of the model φ. A second difficulty is caused by the infinite tangent to the graph of φ at a which means that the contribution h of a single lattice site to the total energy is not bounded above. These features make the statistical mechanical problem considerably more difficult than the one we have considered.

References

[1] R. Bowen and D. Ruelle, The ergodic theory of Axiom A flows, Inventiones Math. 29 181-202 (1975).

[2] G. Gallavotti, Ising model and Bernoulli schemes in one dimension, Commun. Math. Phys. 32 (1973), 183-190.

[3] J. Guckenheimer, A strange strange attractor, in [6], pp. 368-381.

[4] J. Leray, Sur le mouvement d'un liquide visqueax emplissant l'espace, Acta Math. 63 (1934), 193-248.

[5] E. N. Lorenz, Deterministic nonperiodic flow, J. Atmos. Sci. 20 (1963), 130-141.

[6] J. E. Marsden and M. McCracken, The Hopf Bifurcation and its Applications, Applied Mathematical Sciences 19, Springer-Verlag, 1976.

[7] R. M. May, Simple mathematical models with very complicated dynamics, Nature 261 (1976), 459-467.

[8] J. B. McLaughlin and P. C. Martin, Transition to turbulence of a statically stressed fluid system, Phys. Rev. A 12 (1975), 186-203.

[9] D. Ruelle, Statistical mechanics of a one-dimensional lattice gas, Commun. Math. Phys. 9 (1968), 267-278.

[10] D. Ruelle, A measure associated with Axiom A attractors, Amer. Jour. Math., to appear.

[11] S. Smale, Differentiable dynamical systems, Bull. Amer. Math. Soc. 73 (1967), 747-817.

[12] R. F. Williams, The structure of Lorenz attractors, Preprint, Northwestern University (1976).

CENTRO INTERNAZIONALE MATEMATICO ESTIVO
(C.I.M.E.)

MANY PARTICLE COULOMB SYSTEMS

Elliott H. LIEB

Departments of Mathematics and Physics
Princeton University - Princeton, N.J. 08540

Corso tenuto a Bressanone dal 21 giugno al 24 giugno 1976

MANY PARTICLE COULOMB SYSTEMS

Elliott H. Lieb*

Departments of Mathematics and Physics

Princeton University

Princeton, N.J. 08540

Lectures presented at the 1976 session on statistical mechanics of the International Mathematical Summer Center (C.I.M.E.) Bressanone, Italy, June 21-27.

* **Work partially supported by U.S. National Science Foundation grant MCS 75-21684.**

With the introduction of the Schroedinger equation in 1926 it became possible to resolve one of the fundamental paradoxes of the atomic theory of matter (which itself had only become universally accepted a few decades earlier): Why do the electrons not fall into the nucleus? (Jeans, 1915). Following this success, more complicated questions posed themselves. Why is the lowest energy of bulk matter extensive, i.e. why is it proportional to N, the number of particles, instead of to some higher power of N? Next, why do the ordinary laws of thermodynamics hold? Why, in spite of the long range Coulomb force, can a block of matter be broken into two pieces which, after a microscopic separation, are independent of each other?

The aim of these lectures is to answer the above questions in a simple and coherent way. It is a summary of research I have been engaged in for the past few years, and it has been my good fortune to have had the benefit of collaboration with J.L. Lebowitz, B. Simon and W.E. Thirring. Without their insights and stimulation probably none of this could have been carried to fruition.

The accompanying flow chart might be helpful. In section I atoms are shown to be stable because of the Sobolev inequality, not the Heisenberg uncertainty principle. A new inequality related to Sobolev's for functions in the antisymmetric tensor product $L^2(\mathbb{R}^3)^N$ is presented in section II. Thomas-Fermi theory (which was introduced in 1927 just after the Schroedinger equation) is analyzed in section III. This subject is interesting for three reasons: (i) As an application of nonlinear functional analysis; (ii) It turns out that it agrees asymptotically with the Schroedinger equation in a limit in which the nuclear charges go to infinity; (iii) The no-binding theorem of Thomas-Fermi theory, when combined with the inequality of section II, yields a simple proof of the stability of matter. The latter is given in section IV. The first proof of stability is due to Dyson and Lenard in 1967, but the proof in section IV is much simpler. Section V deals with the thermodynamic problem. The difficulty here is not the one of collapse, which was settled in section IV, but the possibility of explosion caused by the long range part of the Coulomb potential. Newton's theorem that a charged sphere behaves from the outside as though all its charge were concentrated at the center, together with some geometric facts about the packing of balls, is used to tame the 1/r potential. Section VI on Hartree-Fock theory is really outside the central theme, but it has been added as a further exercise in functional analysis and because it is, after all, the most common approximation scheme to solve the Schroedinger equation.

Chapters II and IV come from (Lieb-Thirring, 1975), Chapter III from (Lieb-Simon, 1976), Chapter V from (Lieb-Lebowitz, 1972) and Chapter VI from (Lieb-Simon, 1973).

An attempt was made to present the main ideas in as simple and readable a form as possible, and therefore to omit many technical details. There were two reasons for this. The first was to try to make the lectures accessible to physicists as well as to mathematicians. This also creates notational and semantic problems which, it is hoped, have been at least partially resolved. With this aim in mind, I hope the inclusion of such things as an explanation of Young's inequality will be excused. The second reason stems from the belief that if enough hints of a proof are given then a competent analyst would as soon supply the details for himself as read about them. The bibliography is not scholarly, but I believe no theorem has been quoted without proper credit.

I am most grateful to S.B. Treiman who generously devoted much time to reading the manuscript and who made many valuable suggestions to improve its clarity.

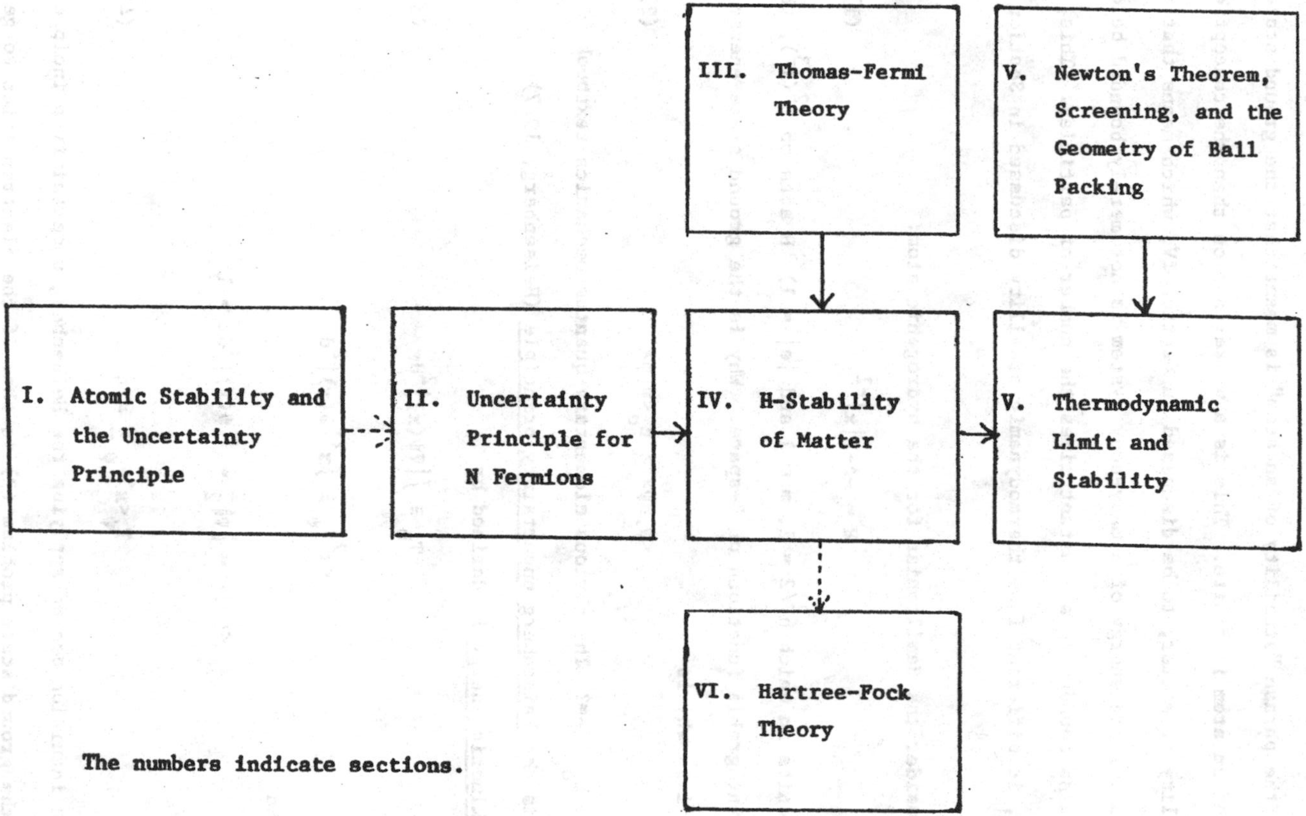

The numbers indicate sections.

I. The Stability of Atoms

By the phrase "stability of an atom" is meant that the ground state energy of an atom is finite. This is a weaker notion than the concept of H-stability of matter, to be discussed in Section IV, which means that the ground state energy of a many-body system is not merely bounded below but is also bounded by a constant times the number of particles. This, in turn, is different from thermodynamic stability discussed in Section V.

Consider the Hamiltonian for the hydrogenic atom:

$$H = -\Delta - Z|x|^{-1} \tag{1}$$

(using units in which $\hbar^2/2 = 1$, $m = 1$ and $|e| = 1$) H acts on $L^2(\mathbb{R}^3)$, the square integrable functions on 3-space. Why is the ground state energy finite, i.e. why is

$$\langle\psi, H\psi\rangle \geq E_0\langle\psi,\psi\rangle \tag{2}$$

for some $E_0 > -\infty$? The obvious elementary quantum mechanics textbook answer is the Heisenberg uncertainty principle (Heisenberg, 1927):
If the kinetic energy is defined by

$$T_\psi \equiv \int|\nabla\psi(x)|^2 dx \tag{3}$$

and if

$$\langle x^2\rangle_\psi \equiv \int x^2|\psi(x)|^2 dx$$

then when

$$\langle\psi,\psi\rangle = \|\psi\|_2^2 = \int|\psi(x)|^2 dx = 1$$

$$T_\psi\langle x^2\rangle_\psi \geq 9/4 \quad . \tag{4}$$

The intuition behind applying the Heisenberg uncertainty principle (4) to the ground state problem (2) is that if the electron tries to get within a distance R of the nucleus, the kinetic energy T_ψ is at least as large as R^{-2}. Consequently $\langle\psi,H\psi\rangle \geq R^{-2}-Z/R$, and this has a minimum $-Z^2/4$ for $R = 2/Z$.

The above argument is false! The Heisenberg uncertainty principle says no such thing, despite the endless invocation of the argument. Consider a ψ consisting of two parts, $\psi = \psi_1+\psi_2$. ψ_1 is a narrow wave packet of radius R centered at the origin with $\int|\psi_1|^2=1/2$. ψ_2 is spherically symmetric and has support in a narrow shell of mean radius L and $\int|\psi_2|^2=1/2$. If L is large then, roughly, $\int x^2|\psi(x)|^2dx \sim L^2/2$ whereas $\int|x|^{-1}|\psi(x)|^2dx \sim 1/2R$. Thus, from (4) we can conclude only that $T_\psi > 9/2L^2$ and hence that $\langle\psi,H\psi\rangle \geq 9/2L^2-Z/2R$. With this wave function, and using only the Heisenberg uncertainty principle, we can make E_o arbitrarily negative by letting $R \to 0$.

A more colorful way to put the situation is this: an electron cannot have both a sharply defined position and momentum. If one is willing to place the electron in two widely separated packets, however, say here and on the moon, then the Heisenberg uncertainty principle alone does not preclude each packet from having a sharp position and momentum.

Thus, while (4) is correct it is a pale reflection of the power of the operator $-\Delta$ to prevent collapse. A better uncertainty principle (i.e. a lower bound for the kinetic energy in terms of some integral of ψ which does not involve derivatives) is needed, one which reflects more accurately the fact that if one tries to compress a wave function anywhere then the kinetic energy will increase. This principle was provided by Sobolev (Sobolev, 1938) and for some unknown reason his inequality, which is simple and goes directly to the heart of the matter, has not made its way into the quantum mechanics textbooks where it belongs. Sobolev's inequality in three dimensions (unlike (4) its form is dimension dependent) is

$$T_\psi = \int|\nabla\psi(x)|^2dx \geq K_s\{\int\rho(x)^3dx\}^{1/3} = K_s\|\rho\|_3 \tag{5}$$

where

$$\rho(x) = |\psi(x)|^2 \tag{6}$$

is the density and

$$K_s = 3(\pi/2)^{4/3} = 5.478$$

is known to be the best possible constant. (5) is non-linear in ρ, but that is unimportant.

A rigorous derivation of (5) would take too long to present but it can be made plausible as follows (Rosen, 1971): K_s is the minimum of

$$K^\psi = \int|\nabla\psi(x)|^2dx \ / \ \{\int|\psi(x)|^6dx\}^{1/3} \quad .$$

Let us accept that a minimizing ψ exists (this is the hard part) and that it satisfies the obvious variational equation

$$-(\Delta\psi)(x) - \alpha\ \psi(x)^5 = 0$$

with $\alpha > 0$. Assuming also that there is a minimizing ψ which is non-negative and spherically symmetric (this can be proved by a rearrangement inequality), one finds by inspection that

$$\psi(x) = (3/\alpha)^{2/3}(1+|x|^2)^{-1/2} \quad .$$

When this is inserted into the expression for K^ψ the result is $K_s = 3(\pi/2)^{4/3}$. The minimizing ψ is not square integrable, but that is of no concern.

Now let us make a simple calculation to show how good (5) really is. For any ψ

$$\langle\psi,H\psi\rangle \geq K_s(\int\rho(x)^3dx)^{1/3} - Z\int|x|^{-1}\rho(x)dx \equiv h(\rho) \tag{7}$$

and hence

$$\langle\psi,H\psi\rangle \geq \min\{h(\rho): \rho(x) \geq 0, \int\rho = 1\} . \tag{8}$$

The latter calculation is trivial (for any potential) since gradients are not involved. One finds that the solution to the variational equation is $\rho(x) = \alpha[|x|^{-1}-R^{-1}]^{1/2}$ for $|x| \leq R$ and $\rho(x) = 0$ for $|x| \geq R$, with $R = K_s \pi^{-4/3}Z^{-1}$. Then

$$h(\rho) = Z^2(\pi/2)^{4/3}/K_s = -(4/3)Z^2Ry \quad .$$

(Recall that one Rydberg = Ry = 1/4 in these units.) Thus, (5) leads easily to the conclusion

$$E_o \geq -(4/3)Z^2Ry \tag{9}$$

and this is an excellent lower bound to the correct $E_o = -Z^2Ry$, especially since no differential equation had to be solved.

In anticipation of later developments a weaker, but also useful, form of (5) can be derived. By Hölder's inequality*

$$\int\rho(x)^{5/3}dx \leq \{\int\rho(x)^3dx\}^{1/3}\{\int\rho(x)dx\}^{2/3} \tag{10}$$

and, since we always take $\int|\psi|^2 = 1$,

$$T_\psi \geq K_s \int\rho(x)^{5/3}dx \quad . \tag{11}$$

---

* Hölder's inequality states that

$$|\int f(x)g(x)dx| \leq \{\int|f(x)|^p dx\}^{1/p} \{\int|g(x)|^q dx\}^{1/q}$$

when $p^{-1}+q^{-1} = 1$ and $p \geq 1$. To obtain (10) take $f = \rho$, $g = \rho^{2/3}$, $p = 3$, $q = 3/2$.

Note that there is now an exponent 1 outside the integral. Although K_s is the best constant in (5) it is not the best constant in (11). Call the latter K_1. K_1 is the minimum of $\int|\nabla\psi(x)|^2dx/\int\rho(x)^{5/3}dx$ subject to $\int\rho(x)dx = 1$. This leads to a non-linear Schroedinger equation whose numerical solution yields (J. Barnes, private communication)

$$K_1 \approx 9.578 \quad .$$

In any event

$$K_1 > K^c \equiv (3/5)(6\pi^2)^{2/3} \approx 9.116$$

and hence

$$T_\psi \geq K^c \int\rho(x)^{5/3}dx \text{ when } \int|\psi|^2 = 1 \ . \tag{12}$$

K^c is much bigger than K_s; it is the classical value, and will be encountered again in section II and in section III where its significance will be clarified.

We can repeat the minimization calculation analogous to (8) using the bound (12) and the functional

$$h^c(\rho) = K^c\int\rho(x)^{5/3}dx - Z\int|x|^{-1}\rho(x)dx \quad .$$

(We could, of course, use the better constant K_1.) This time

$$\rho(x) = \{(3/5)(Z/K^c)(|x|^{-1}-R^{-1})\}^{3/2} \tag{13}$$

for $|x| \leq R$. R is determined by $\int\rho = 1$ and one finds that $R = (K^c/Z)(4/\pi^2)^{2/3}$ and

$$E_o \geq -(9Z^2/5K^c)(\pi^2/4)^{2/3} = -3^{1/3}\, Z^2 \text{ Ry} \ . \tag{14}$$

$3^{1/3}$ is only 8.2% greater than 4/3.

The Sobolev inequality (5) or its variant (12) is, for our purposes, a much better uncertainty principle than Heisenberg's - indeed it is also fairly accurate. We now want to extend (12) to the N-particle case in order to establish the stability of bulk matter. The important new fact that will be invoked is that the N particles are fermions; that is to say the N-particle wave function is an antisymmetric function of the N-space, spin variables.

II. Extension of the Uncertainty Principle to Many Fermions

A well known elementary calculation is that of the lowest kinetic energy, T^V, of N *fermions* in a cubic box of volume V. For large N one finds that

$$T^V \simeq q^{-2/3} K^c V \rho^{5/3} \tag{15}$$

where $\rho = N/V$ and q is the number of spin states available to each particle (q=2 for electrons). (15) is obtained by merely adding up the N/q lowest eigenvalues of $-\Delta$ with Dirichlet ($\psi=0$) boundary conditions on the walls of the box. The important feature of (15) is that it is proportional to $N^{5/3}$ instead of N, as would be the case if the particles were not fermions. The extra factor $N^{2/3}$ is essential for the stability of matter; if electrons were bosons matter would not be stable.

(15) suggests that (12), with a factor $q^{-2/3}$, ought to extend to the N-particle case if $\rho(x)$ is interpreted properly. The idea is old, going back to Lenz (Lenz, 1932) who got it from Thomas-Fermi theory. The proof that something like (12) is not only an approximation but is also a lower bound is new.

To say that the N particles are *fermions* with q spin states means that the N-particle wave function $\psi\,(x_1,\ldots,x_N;\,\sigma_1,\ldots,\sigma_N)$ defined for $x_i \in \mathbb{R}^3$ and $\sigma_i \in \{1,2,\ldots,q\}$ is *antisymmetric* in the pairs (x_i,σ_i). The norm is given by

$$\langle\psi,\psi\rangle = \sum_{\sigma_i=1}^{q} \int |\psi(x_1,\ldots,x_N;\sigma_1,\ldots,\sigma_N)|^2 dx_1\ldots dx_N \ .$$

Define

$$T_\psi = N \sum_{\sigma_i=1}^{q} \int |\nabla_1\psi(x_1,\ldots,x_N;\sigma_1,\ldots,\sigma_N)|^2 dx_1\ldots dx_N \tag{16}$$

to be the usual kinetic energy of ψ and define

$$\rho_\psi(x) = N \sum_{\sigma_i=1}^{q} \int |\psi(x,x_2,\ldots,x_N;\sigma_1,\ldots,\sigma_N)|^2 dx_2\ldots dx_N \tag{17}$$

to be the single particle density, i.e. the probability of finding a particle at x. The analogue of (12) is the following.

<u>Theorem 1</u>. <u>If</u> $\langle\psi,\psi\rangle = 1$ <u>then</u>

$$T_\psi \geq (4\pi q)^{-2/3} K^c \int \rho_\psi(x)^{5/3} dx \quad . \tag{18}$$

Apart from the annoying factor $(4\pi)^{-2/3} \approx 0.185$, (18) says that the intuition behind considering (15) as a lower bound is correct. We believe that $(4\pi)^{-2/3}$ does not belong in (18) and hope to eliminate it someday. Recent work (Lieb, 1976) has improved the constant by a factor $(1.83)^{2/3} = 1.496$, so we are now off from the conjectured constant $q^{-2/3}K^c$ only by the factor 0.277.

The proof of Theorem 1 is not long but it is slightly tricky. It is necessary first to investigate the negative eigenvalues of a one-particle Schroedinger equation when the potential is non-positive.

<u>Theorem 2</u>. <u>Let</u> $V(x) \leq 0$ <u>be a potential for the one-particle, three dimensional Schroedinger operator</u> $H = -\Delta+V(x)$ <u>on</u> $L^2(\mathbb{R}^3)$. <u>For</u> $E \leq 0$ <u>let</u> $N_E(V)$ <u>be the number of eigenstates of</u> H <u>with energies</u> $\leq$ E. <u>Then</u>

$$N_E(V) \leq (4\pi)^{-1}(2|E|)^{-1/2} \int |V(x)-E/2|_-^2 \, dx \tag{19}$$

<u>where</u> $|f(x)|_- = |f(x)|$ <u>if</u> $f(x) \leq 0$ <u>and</u> $|f(x)|_- = 0$ <u>otherwise</u>.

<u>Corollary</u>. <u>If</u> $e_1 \leq e_2 \leq \ldots \leq 0$ <u>are the negative eigenvalues of</u> H <u>(if any) then</u>

$$\sum_{j=1} |e_j| \leq (4\pi/15\pi^2) \int |V(x)|^{5/2} dx \quad . \tag{20}$$

Proof. $\Sigma|e_j| = \int_0^\infty N_{-\alpha}(V)d\alpha$. Insert (19) and do the α integration first and then the x integration. The result is (20). ▨

We believe the factor (4π) does not belong in (20).

Proof of Theorem 2. From the Schroedinger equation $H\psi = e\psi$ it is easy to deduce that $N_E(V)$ is equal to the number of eigenvalues which are ≥ 1 of the positive definite Birman-Schwinger operator (Birman, 1961, Schwinger, 1961)

$$B_E(V) = |V|^{1/2} (-\Delta-E)^{-1} |V|^{1/2}. \tag{21}$$

Essentially (21) comes from the fact that if $H\psi = e\psi$ then $(-\Delta-e)\psi = |V|\psi$. If one defines $|V|^{1/2}\psi \equiv \phi$, then $B_e\phi = \phi$. Thus B_e has an eigenvalue 1 when e is an eigenvalue. However, B_E is a compact positive semi-definite operator on $L^2(\mathbb{R}^3)$ for $E < 0$ and, as an operator, B_E is monotone decreasing in E. Thus, if B_E has k eigenvalues ≥ 1, there exist k numbers $e_1 \leq e_2 \leq \ldots \leq e_k \leq E$ such that B_{e_j} has eigenvalue 1.

Consequently $N_E(V) \leq \mathrm{Tr}\, B_E(V)^2$. On the other hand $N_E(V) \leq N_{E/2}(-|V-E/2|_-)$ by the variational principle (draw a graph of $V(x)-E/2$). Thus, since $B_E(V)$ has a kernel $B_E(x,y) = |V(x)|^{1/2} \exp\{-|E|^{1/2}|x-y|\}[4\pi|x-y|]^{-1}|V(y)|^{1/2}$.

$$\begin{aligned} N_E(V) &\leq \mathrm{Tr}\, B_{E/2}\,(-|V-E/2|_-)^2 \\ &= (4\pi)^{-2} \iint dxdy\,|V(x)-E/2|_-\;|V(y)-E/2|_- \\ &\quad\cdot \exp\{-(2|E|)^{1/2}|x-y|\}\;|x-y|^{-2}\;. \end{aligned} \tag{22}$$

(19) results from applying Young's inequality* to (22). Alternatively, one can do the convolution integral by Fourier transforms and note that the Fourier transform of the last factor has a maximum at $p = 0$ where it is $4\pi(2|E|)^{-1/2}$. ▮

Using (20), which is a statement about the energy levels of a single particle Hamiltonian we can, surprisingly, prove Theorem 1 which refers to the kinetic energy of N fermions.

<u>Proof of Theorem 1</u>. ψ and hence $\rho_\psi(x)$ are given. Consider the non-positive single particle potential $V(x) \equiv -\alpha\rho_\psi(x)^{2/3}$ where α is given by $(2/3\pi)\, q\, \alpha^{3/2} = 1$. Next consider the following N-particle Hamiltonian

$$\tilde{H}_N = \sum_{i=1}^{N} h_i; \qquad h_i = -\Delta_i + V(x_i)$$

on $L^2(\mathbb{R}^3;\mathbb{C}^q)^N$. If E_o is the <u>fermion</u> ground state energy of $\tilde{H}_N$ we have that $E_o \geq q\,\Sigma e_j$, where the e_j are the negative eigenvalues of the single particle Hamiltonian h. (We merely fill the lowest negative energy levels q times until the N particles are accounted for; if there are k such levels and if $N < kq$ then $E_o > q\,\Sigma e_j$. If $N > kq$, the surplus particles can be placed in wave packets far away from the origin with arbitrarily small kinetic energy.) On the other hand, $E_o \leq \langle\psi, \tilde{H}_N \psi\rangle = T_\psi - \alpha\int\rho_\psi(x)^{5/3}dx$ by the variational principle. If these two inequalities are combined together with (20), which says that $\Sigma\, e_j \geq -(4/15\pi)\alpha^{5/2}\int\rho_\psi(x)^{5/3}dx$, then (18) is the result. ▮

* Young's inequality states that

$$\left|\iint f(x)g(x-y)h(y)\,dx\,dy\right| \leq \{\int|f(x)|^p dx\}^{1/p}\{\int|g(x)|^q\}^{1/q}\{\int|h(x)|^r dx\}^{1/r}$$

when $p^{-1}+q^{-1}+r^{-1} = 2$ and $p,q,r \geq 1$. For (22) take $p=r=2$ and $q=1$.

It might not be too much out of place to explain at this point why K^c is called the classical constant. The name does not stem from its antiquity, as in the ideal gas kinetic energy (15), but rather from classical mechanics -- more precisely the semiclassical approximation to quantum mechanics. This intuitive idea is valuable.

As the proof of Theorem 1 shows, the constant in (18) for T_ψ is simply related to the constant in (20) for the sum of the eigenvalues. The point is that the semiclassical approximation to this sum is

$$\Sigma|e_j| \approx (15\pi^2)^{-1} \int |V(x)|^{5/2}\, dx$$

and this, in turn, would yield (18) without the $(4\pi)^{-2/3}$ factor. The semiclassical approximation is obtained by saying that a region of volume $(2\pi)^3$ in the 6-dimensional phase space (p,x) can accommodate one eigenstate. Hence, integrating over the set $\theta(H)$, in which $H(p,x) = p^2+V(x)$ is negative,

$$\begin{aligned}\Sigma e_j &\approx (2\pi)^{-3} \int\int_{\theta(H)} dxdp\, \{p^2+V(x)\}\\ &= (2\pi)^{-3} \int dx\ \ 4\pi \int_0^{\sqrt{|V(x)|}} p^2dp\{p^2+V(x)\}\\ &= -(15\pi^2)^{-1} \int |V(x)|^{5/2}\, dx\,.\end{aligned}$$

If a coupling constant g is introduced, and if V is replaced by gV, then it is a theorem that the semiclassical approximation is asymptotically exact as $g \to \infty$ for any V in $L^{5/2}(\mathbb{R}^3)$.

Theorem 1 gives a lower bound to the kinetic energy of fermions which is crucial for the H-stability of matter as developed in Section IV. To appreciate the significance of Theorem 1 it should be compared with the one-particle Sobolev bound (12). Suppose that $\rho(x) = 0$ outside some fixed domain, Ω, of volume V. Then since

$$\int_\Omega \rho(x)^{5/3} dx \geq \{\int_\Omega \rho(x) dx\}^{5/3} \{\int_\Omega 1\}^{-2/3} = N^{5/3} V^{-2/3}$$

by Hölder's inequality, one sees that T_ψ grows at least as fast as $N^{5/3}$. Using (12) alone, one would only be able to conclude that T_ψ grows as N. This distinction stems from the Pauli principle, i.e. the antisymmetric nature of the N-particle wave function. As we shall see, this $N^{5/3}$ growth is essential for the stability of matter because without it the ground state energy of N particles with Coulomb forces would grow at least as fast as $-N^{7/5}$ instead of -N.

The Fermi pressure is needed to prevent a collapse, but to learn how to exploit it we must first turn to another chapter in the theory of Coulomb systems, namely Thomas-Fermi theory.

III. Thomas-Fermi Theory

The statistical theory of atoms and molecules was invented independently by Thomas and Fermi (Thomas, 1927, Fermi, 1927). For many years the TF theory was regarded as an uncertain approximation to the N-particle Schroedinger equation and much effort was devoted to trying to determine its validity (e.g. Gombás, 1949). It was eventually noticed numerically (Sheldon, 1955) that molecules did not appear to bind in this theory, and then Teller (Teller, 1962) proved this to be a general theorem.

It is now understood that TF theory is really a large Z theory (Lieb-Simon, 1976); to be precise it is exact in the limit $Z \to \infty$. For finite Z, TF theory is qualitatively correct in that it adequately describes the bulk of an atom or molecule. It is not precise enough to give binding. Indeed, it should not do so because binding in TF theory would imply that the cores of atoms bind, and this does not happen. Atomic binding is a fine quantum effect. Nevertheless, TF theory deserves to be well understood because it is exact in a limit; the TF theory is to the many electron system as the hydrogen atom is to the few electron system. For this reason the main features of the theory are presented here, mostly without proof.

A second reason for our interest in TF theory is this: in the next section the problem of the H-stability of matter will be reduced to a TF problem. The knowledge that TF theory is H-stable (this is a corollary of the no binding theorem) will enable us to conclude that the true quantum system is H-stable.

The Hamiltonian for N electrons with k static nuclei of charges $z_i > 0$ and locations R_i is

$$H_N = \sum_{i=1}^{N} -\Delta_i - V(x_i) + \sum_{1\le i<j\le N} |x_i - x_j|^{-1} + U(\{z_j, R_j\}_{j=1}^{k}) \tag{23}$$

where

$$V(x) = \sum_{j=1}^{k} z_j |x - R_j|^{-1} \tag{24a}$$

and

$$U(\{z_j, R_j\}_{j=1}^{k}) = \sum_{1\le i<j\le k} z_i z_j |R_i - R_j|^{-1} . \tag{24b}$$

The nuclear-nuclear repulsion U is, of course, a constant term in H_N but it is included for two reasons:

(i) We wish to consider the dependence on the R_i of

$$E_N^Q(\{z_j, R_j\}_{j=1}^{k}) \equiv \text{the ground state energy of } H_N . \tag{25}$$

(ii) Without U the energy will not be bounded by N.

The nuclear kinetic energy is not included in H_N. For the H-stability problem we are only interested in finding a lower bound to E_N^Q, and the nuclear kinetic energy adds a positive term. In other words, $\inf_{\{R_i\}} E_N^Q(\{z_j, R_j\}_{j=1}^{k})$ is smaller than the ground state energy of the true Hamiltonian (defined in (58)) in which the nuclear kinetic energy is included. Later on when we do the proper thermodynamics of the whole system we shall have to include the nuclear kinetic energy.

The problem of estimating E_N^Q is as old as the Schroedinger equation. The TF theory, as interpreted by Lenz (Lenz, 1932), reads as follows: For fermions having q spin states (q= 2 for electrons) define the <u>TF energy functional</u>:

$$\mathcal{E}(\rho) = q^{-2/3} K^c \int \rho(x)^{5/3} - \int V(x)\rho(x) + \frac{1}{2}\iint \rho(x)\rho(y)|x-y|^{-1} dx dy + U(\{z_j, R_j\}_{j=1}^{k}) \tag{26}$$

for <u>non-negative</u> functions $\rho(x)$. Then for $\lambda \ge 0$

$$E_\lambda^{TF} \equiv \inf\{\mathcal{E}(\rho): \int\rho(x)dx = \lambda\} \tag{27}$$

is the TF energy for λ electrons (λ need not be an integer, of course). When $\lambda = N$ the minimizing ρ is supposed to approximate the ρ_ψ given by (17), wherein ψ is the true ground state wave function, and E_N^{TF} is supposed to approximate E_N^Q.

The second and fourth terms on the right side of (26) are exact but the first and third are not. The first is to some extent justified by the kinetic energy inequality, Theorem 1; the third term will be discussed later. In any event, (26) and (27) <u>define</u> TF theory.

It would be too much to try to reproduce here the details of our analysis of TF theory. A short summary of some of the main theorems will have to suffice.

The first question is whether or not E_λ^{TF} (which, by simple estimates using Young's and Hölder's inequalities can be shown to be finite for all λ) is a minimum as distinct from merely an infimum. The distinction is crucial because the <u>TF equation</u> (the Euler-Lagrange equation for (26) and (27)):

$$(5/3)K^c q^{-2/3}\rho^{2/3}(x) = \max\{\phi(x)-\mu, 0\} \tag{28}$$

with

$$\phi(x) = V(x) - \int\rho(y)|x-y|^{-1}dy \tag{29}$$

has a solution with $\int\rho=\lambda$ if and only if there is a minimizing ρ for E_λ^{TF}. The basic theorem is

<u>Theorem 3</u>. <u>If</u> $\lambda \leq Z \equiv \sum_{j=1}^{k} z_j$ <u>then</u>

(i) <u>$\mathcal{E}(\rho)$ has a minimum on the set</u> $\int\rho(x)dx = \lambda$.

(ii) This minimizing ρ (call it ρ_λ^{TF}) is unique and satisfies (28) and (29). μ is non-negative, and $-\mu$ is the chemical potential, i.e.

$$-\mu = dE_\lambda^{TF}/d\lambda \quad . \tag{30}$$

(iii) There is no other solution to (28) and (29) (for any μ) with $\int\rho=\lambda$ other than ρ_λ^{TF}.

(iv) When $\lambda = Z$, $\mu = 0$. Otherwise $\mu > 0$, i.e. E_λ^{TF} is strictly decreasing in λ.

(v) As λ varies from 0 to Z, μ varies continuously from $+\infty$ to 0.

(vi) μ is a convex, decreasing function of λ.

(vii) $\phi_\lambda^{TF}(x) > 0$ for all x and λ. Hence when $\lambda = Z$

$$(5/3)K^c\, q^{-2/3}\, \rho_Z^{TF}(x)^{2/3} = \phi_Z^{TF}(x) \ .$$

If $\lambda > Z$ then $E^{TF}(\lambda)$ is not a minimum and (28) and (29) have no solution with $\int\rho = \lambda$. Negative ions do not exist in TF theory. Nevertheless, E_λ^{TF} exists and $E_\lambda^{TF} = E_Z^{TF}$ for $\lambda \geq Z$.

The proof of Theorem 3 is an exercise in functional analysis. Basically, one first shows that $\mathcal{E}(\rho)$ is bounded below so that E_λ^{TF} exists. The Banach-Alaoglu theorem is used to find an $L^{5/3}$ weakly convergent sequence of ρ's such that $\mathcal{E}(\rho)$ converges to E_λ^{TF}. Then one notes that $\mathcal{E}(\rho)$ is weakly lower semicontinuous so that a minimizing ρ exists. The uniqueness comes from an important property of $\mathcal{E}(\rho)$, namely that it is convex. This also implies that the minimizing ρ satisfies $\int\rho = \lambda$. A major point to notice is that a solution of the TF equation is obtained as a byproduct of minimizing $\mathcal{E}(\rho)$; a direct proof that the TF equation has a solution would be very complicated.

Apart from the details presented in Theorem 3, the main point is that TF theory is well defined. In particular the density ρ is unique - a state of affairs in marked contrast to that of Hartree-Fock theory.

The TF density ρ_λ^{TF} has the following properties:

Theorem 4. If $\lambda \leq Z$ then

(i) $(5/3)K^c\, q^{-2/3}\, \rho_\lambda^{TF}(x)^{2/3} \sim z_i|x-R_i|^{-1}$ (31)

near each R_i.

(ii) In the neutral case, $\lambda = Z = \sum_{j=1}^{k} z_j$,

$$|x|^6\, \rho_Z^{TF}(x) \rightarrow (3/\pi)^3[5/3\; K^c\, q^{-2/3}]^3 \tag{32}$$

as $|x| \rightarrow \infty$, irrespective of the distribution of the nuclei.

(iii) $\phi_\lambda^{TF}(x)$ and $\rho_\lambda^{TF}(x)$ are real analytic in x away from all the R_i, on all of 3-space in the neutral case and on $\{x\colon \phi_\lambda^{TF}(x) > \mu\}$ in the positive ionic case.

(32) is especially remarkable: at large distances one loses all knowledge of the nuclear charges and configuration. Property (i) recalls the singularity found in the minimization of $h^c(\rho)$ (see (13)).

(31) can be seen from (28) and (29) by inspection. (32) is more subtle but it is consistent with the observation that (28) and (29) can be rewritten (when $\mu=0$) as

$$-(4\pi)^{-1}\Delta\, \phi_Z^{TF}(x) = -\{(3/5)q^{2/3}\, \phi_Z^{TF}(x)/K^c\}^{3/2}$$

away from the R_i. If it is assumed that $\phi_Z^{TF}(x)$ goes to zero as a power of $|x|$ then (32) follows. This observation was first made by Sommerfeld (Sommerfeld, 1932). The proof that a power law falloff actually occurs is somewhat subtle and involves potential theoretic ideas such as that used in the proof of Lemma 8.

As pointed out earlier, the connection between TF theory and the Schroedinger equation is best seen in the limit $Z \to \infty$. Let the number, k, of nuclei be held fixed, but let $N \to \infty$ and $z_i \to \infty$ in such a way that the degree of ionization N/Z is constant, where $Z = \sum_{j=1}^{k} z_j$. To this end we make the following definition: Fix $\{z_j, R_j\}_{j=1}^{k}$ and λ. It is not necessary to assume that $\lambda \leq Z$. For each N = 1,2,... define a_N by $\lambda a_N = N$. In H_N (23) replace z_j by $z_j a_N$ and R_j by $R_j a_N^{-1/3}$. This means that the nuclei come together as $N \to \infty$. If they stay at fixed positions then that is equivalent, in the limit, to isolated atoms, i.e. it is equivalent to starting with all the nuclei infinitely far from each other. Finally for the nuclear configuration $\{a_N z_j, a_N^{-1/3} R_j\}_{j=1}^{k}$ let ψ_N be the ground state wave function, E_N^Q the ground state energy, and $\rho_N^Q(x)$ be the single particle density as defined by (17). [Note: If E_N^Q is degenerate, ψ can be <u>any</u> ground state wave function as far as Theorem 5 is concerned. If E_N^Q is not an eigenvalue, but merely inf spec H_N, then it is possible to define an approximating sequence ψ_N, with ρ_N^Q still given by (17), in such a way that Theorem 5 holds. We omit the details of this construction here.]

It is important to note that there is a simple and obvious scaling relation for TF theory, namely

$$E_{\lambda a}^{TF}(\{az_j, a^{-1/3}R_j\}_{j=1}^{k}) = a^{7/3}\, E_{\lambda}^{TF}(\{z_j, R_j\}_{j=1}^{k}) \tag{33}$$

and

$$\rho_{\lambda a}^{TF}(a^{-1/3}x) = a^2\, \rho_{\lambda}^{TF}(x) \tag{34}$$

for any $a \geq 0$. Hence, for the above sequence of systems parametrized by a_N,

$$a_N^{-7/3} E_N^{TF}(\{a_N z_j, a_N^{-1/3} R_j\}) = E_\lambda^{TF}(\{z_j, R_j\}_{j=1}^k) \tag{35}$$

$$a_N^{-2} \rho_N^{TF}(a_N^{-1/3} x) = \rho_\lambda^{TF}(x) \tag{36}$$

for all N.

If, on the other hand, the nuclei are held fixed then one can prove that

$$\lim_{N\to\infty} a_N^{-7/3} E_N^{TF}(\{a_N z_j, R_j\}) = \sum_{j=1}^k E_{\lambda_j}^{TF}(z_j) \tag{37}$$

where $E_{\lambda_j}^{TF}(z)$ is the energy of an isolated atom of nuclear charge z. The λ_j are determined by the condition that $\sum_{j=1}^k \lambda_j = \lambda$ if $\lambda \leq Z$ (otherwise $\sum_{j=1}^k \lambda_j = Z$) and that the chemical potentials of the k atoms are all the same. Another way to say this is that the λ_j minimize the right side of (37). With the nuclei <u>fixed</u>, the analogue of (36) is

$$\lim_{N\to\infty} a_N^{-2} \rho_N^{TF} (a^{-1/3}(x-R_j)) = \rho_{\lambda_j}^{TF}(x) \quad . \tag{38}$$

The right side of (38) is the ρ for a single atom of nuclear charge z and electron charge λ_j. (37) and (38) are a precise statement of the fact that isolated atoms result from fixing the R_j.

The TF energy for an isolated, <u>neutral</u> atom of nuclear charge Z is found numerically to be

$$E_Z^{TF} = -(2.21)\, q^{2/3} (K^c)^{-1} Z^{7/3} \quad . \tag{39}$$

For future use note that E_Z^{TF} is proportional to $1/K^c$. Thus, if one considers a TF theory with K^c replaced by some other constant $\alpha > 0$, as will be necessary in Section IV, then (39) is correct if K^c is replaced by α.

Theorem 5. With $a_N = N/\lambda$ and $\{z_j, R_j\}_{j=1}^k$ fixed

(i) $a_N^{-7/3} E_N^Q(\{a_N z_j, a_N^{-1/3} R_j\})$ has a limit as $N \to \infty$.

(ii) This limit is $E_\lambda^{TF}(\{z_j, R_j\}_{j=1}^k)$.

(iii) $a_N^{-7/3} E_N^Q(\{a_N z_j, R_j\})$ has a limit as $N \to \infty$. This limit is the right side of (37).

(iv) $a_N^{-2}\, \rho_N^Q(a_N^{-1/3} x; \{a_N z_j, a_N^{-1/3} R_j\})$ also has a limit as $N \to \infty$. If $\lambda \le Z = \sum_{j=1}^k z_j$, this limit is $\rho_\lambda^{TF}(x)$ and the convergence is in weak $L^1(\mathbb{R}^3)$. If $\lambda > Z$, the limit is $\rho_Z^{TF}(x)$ in weak $L^1_{loc}(\mathbb{R}^3)$.

(v) For fixed nuclei, $a_N^{-2}\, \rho_N^Q(a_N^{-1/3}(x - R_j); \{a_N z_j, R_j\})$ has a limit (in the same sense as (iv)) which is the right side of (38).

The proof of Theorem 5 does not use anything introduced so far. It is complicated, but elementary. One partitions 3-space into boxes with sides of order $Z^{-1/3}$. In each box the potential is replaced by its maximum (resp. minimum) and one obtains an upper (resp. lower) bound to E_N^Q by imposing Dirichlet ($\psi=0$) (resp. Neumann ($\nabla\psi=0$)) boundary conditions on the boxes. The upper bound is essentially a Hartree-Fock calculation. The $-r^{-1}$ singularity near the nuclei poses a problem for the lower bound, and it is tamed by exploiting the concept of angular momentum barrier.

What Theorem 5 says, first of all, is that the true quantum energy has a limit on the order of $Z^{7/3}$ when the ratio of electron to nuclear charge is held fixed. Second, this limit is given correctly by TF theory as is shown in (35). The requirement that the nuclei move together as $Z^{-1/3}$ should be regarded as a refinement rather than as a drawback, for if the nuclei are fixed a limit also exists but it is an uninteresting one of isolated atoms.

Theorem 5 also says that the density, ρ_N^Q, is proportional to Z^2 and has a scale length proportional to $Z^{-1/3}$. If $\lambda > Z$, Theorem 5 states that the surplus charge moves off to infinity and the result is an isolated molecule. This means that large atoms or molecules cannot have a <u>negative</u> ionization proportional to the total nuclear charge; at best they can have a negative ionization which is a vanishingly small fraction of the total charge. This result is physically obvious for electrostatic reasons, but it is nice to have a proof of it.

Theorem 5 also resolves certain "anomalies" of TF theory which are:

(a) In real atoms or molecules the electron density falls off exponentially, while in TF theory (Theorem 4) the density falls off as $|x|^{-6}$.

(b) The TF atom shrinks in size as $Z^{-1/3}$ (cf. (36)) while real large atoms have roughly constant size.

(c) In TF theory there is no molecular binding, as we shall show next, but there is binding for real molecules.

(d) In real molecules the electron density is finite at the nuclei, but in TF theory it goes to infinity as $z_j|x-R_j|^{-3/2}$ (Theorem 4).

As Theorem 5 shows, TF theory is really a theory of heavy atoms or molecules. A large atom looks like a stellar galaxy, poetically speaking. It has a core which shrinks as $Z^{-1/3}$ and which contains most of the electrons. The density (on a scale of Z^2) is not finite at the nucleus because, as the simplest Bohr theory shows, the S-wave electrons have a density proportional to Z^3 which is infinite on a scale of Z^2. Outside the core is a mantle in which the density is proportional to (cf. Theorem 4) $(3/\pi)^3[(5/3)K^c 2^{-2/3}]^3\, Z^2/(Z^{1/3}|x|)^6$ which is <u>independent</u> of Z!

This density is correct to infinite distances on a length scale $Z^{-1/3}$. The core and the mantle contain 100% of the electrons as $Z\to\infty$. The third region is a transition region to the outer shell and while it may contain many electrons, it contains only a vanishingly small fraction of them. The fourth region is the outer shell in which chemistry and binding takes place. TF theory has nothing to say about this region. The fifth region is the one in which the density drops off exponentially.

Thus, TF theory deals only with the core and the mantle in which the bulk of the energy and the electrons reside. There ought not to be binding in TF theory, and indeed there is none, because TF energies are proportional to $Z^{7/3}$ and binding energies are of order one. The binding occurs in the fourth layer.

An important question is what is the next term in the energy beyond the $Z^{7/3}$ term of TF theory. Several corrections have been proposed: (e.g. Dirac, 1930, Von Weizsäcker, 1935, Kirzhnits, 1957, Kompaneets and Pavlovskii, 1956, Scott, 1952). With the exception of the last, all these corrections are of order $Z^{5/3}$. Scott (as late as 1952!) said there should be a $Z^{6/3}$ correction because TF theory is not able to treat correctly the innermost core electrons. Recall that in Bohr theory each inner electron alone has an energy proportional to Z^2. As these inner electrons are unscreened, their energies should be independent of the presence or absence of the electron-electron repulsion. In other words, the Z^2 correction for a molecule should be precisely a sum of corrections, one for each atom. The atomic correction should be the difference between the Bohr energy and the $Z^{7/3}$ TF energy for an atom in which the electron-electron repulsion is <u>neglected</u>. We already calculated the TF energy for such an "atom" in (14) (put Z=1 there and then use scaling; also replace K^c by $q^{-2/3}K^c$). Thus, for a <u>neutral</u> atom <u>without</u> electron-electron repulsion

$$\tilde{E}_Z^{TF} = -(3^{1/3}/4)q^{2/3}Z^{7/3} \quad . \tag{40}$$

For the Bohr atom, each shell of energy $-Z^2/4n^2$ has n^2 states, so

$$Z/q = N/q = \sum_{n=1}^{L} n^2+(L+1)^2\phi = L^3/3+L^2/2+L/6+(L+1)^2\phi$$

with $0 \le \phi \le 1$ being the fraction of the (L+1)th shell that is filled. One finds $L \approx (3Z/q)^{1/3} - 1/2 - \phi + o(1)$ and

$$E_Z^{Bohr} = -(Z^2/4)q\{\phi+\sum_{n=1}^{L} 1\} \approx \tilde{E}_Z^{TF} + (q/8)Z^2 \quad .$$

Thus, to the next order, the energy should be

$$E_N^Q(\{z_j,R_j\}_{j=1}^k) = E_N^{TF}(\{z_j,R_j\}_{j=1}^k) + \frac{1}{4}\sum_{j=1}^{k} z_j^2 + \text{lower order}, \tag{41}$$

since q = 2 for electrons. Note that $E_N^{TF} \sim q^{2/3}$ while the Scott correction ~ q.

It is remarkable that (41) gives a <u>precise conjecture</u> about the next correction. It is simple to understand physically, yet we do not have the means to prove it.

The third main fact about TF theory is that there is no binding. This was proved by Teller in 1962. Considering the effort that went into the study of TF theory since its inception in 1927, it is remarkable that the no-binding phenomenon was not seriously noticed until the computer study of Sheldon in 1955. Teller's original proof involved some questionable manipulation with δ-functions and for that reason his result was questioned. His ideas were basically right, however, and we have made them rigorous.

Theorem 6 (no binding). If there are at least two nuclei, write the nuclear attraction $V(x) = \sum_{j=1}^{k} z_j |x-R_j|^{-1}$ as the sum of two pieces, $V = V^1+V^2$ where $V^1(x) = \sum_{j=1}^{m} z_j |x-R_j|^{-1}$ and $1 \le m < k$. Let $E_\lambda^{TF,1}$ be the TF energy for the nuclei $1,\dots,m$ (with $U = \sum_{1\le i<j\le m} z_i z_j |R_i-R_j|^{-1}$, of course) and let $E_\lambda^{TF,2}$ be the same for the nuclei $m+1,\dots,k$. Given λ, let $\lambda_1 \ge 0$ and $\lambda_2 = \lambda-\lambda_1 \ge 0$ be chosen to minimize the sum of the energies of the separate molecules, i.e. $E_{\lambda_1}^{TF,1} + E_{\lambda_2}^{TF,2}$. (If $\lambda = Z = \sum_{j=1}^{k} z_j$ then by Theorem 3, $\lambda_1 = \sum_{j=1}^{m} z_j$.) Then

$$E_\lambda^{TF} \ge E_{\lambda_1}^{TF,1} + E_{\lambda_2}^{TF,2}. \tag{42}$$

Since the right side of (42) is the energy of two widely separated molecules, with the relative nuclear positions unchanged within each molecule, Theorem 6 says that the TF energy is unstable under every decomposition of the big molecule into smaller molecules. In particular, a molecule is unstable under decomposition into isolated atoms, and Theorem 9 is a simple consequence of this fact. One would suppose that if λ and the z_j are fixed, but the R_j are replaced by αR_j then

$$E_\lambda^{TF}(\{z_j, \alpha R_j\}_{j=1}^{k}) \text{ is monotone decreasing in } \alpha.$$

In other words, the "pressure" is always positive. This is an unproved conjecture, but it has been proved (Balàzs, 1967) in the case $k = 2$ and $z_1 = z_2$.

An interesting side remark is

Theorem 7. If the TF energy (26), (27) is redefined by excluding the repulsion term U in (26), then the inequality in (42) is reversed.

Thus, the nuclear repulsion is essential for the no-binding Theorem 6.

Another useful fact for some further developments of the theory, especially the TF theory of solids and the TF theory of screening (Lieb-Simon, 1976) is the following lemma (also due to Teller) which is used to prove the main no binding Theorem 6.

Lemma 8. Fix $\{R_j\}_{j=1}^k$ and fix $\mu \geq 0$ in the TF equation (28) but not $\{z_j\}_{j=1}^k$. (This means that as the z_j's are varied λ will vary, but always $0 \leq \lambda \leq Z = \Sigma z_j$. If $\mu = 0$ then $\lambda = Z$ always.) If $\{z_j^1\}_{j=1}^k$ and $\{z_j^2\}_{j=1}^k$ are two sets of z's such that

$$z_j^1 \leq z_j^2 \text{ all } j, \text{ and } z_1^1 < z_1^2$$

and if λ_1 and λ_2 are the corresponding λ's for the two sets, then for all x

$$\phi_{\lambda_1}^{TF}(x) \leq \phi_{\lambda_2}^{TF}(x)$$

and hence

$$\rho_{\lambda_1}^{TF}(x) \leq \rho_{\lambda_2}^{TF}(x) .$$

There is strict inequality when $\mu = 0$. In short, increasing some z_j increases the density everywhere, not just on the average.

The proof of Lemma 8 involves a beautifully simple potential theoretic argument which we cannot resist giving.

<u>Proof of Lemma 8.</u> We want to prove $\phi_1^{TF}(x) \leq \phi_2^{TF}(x)$ for all x, and will content ourselves here with proving only $\leq$ when $\mu=0$. Let $B = \{x: \phi_1^{TF}(x) > \phi_2^{TF}(x)\}$. B is an open set and B does not contain any R_i for which $z_i^1 < z_i^2$ by the TF equation (29). Let $\psi(x) = \phi_1^{TF}(x) - \phi_2^{TF}(x)$. If $x \in B$ then $\psi(x) > 0$ and, by (28), $\rho_1^{TF}(x) > \rho_2^{TF}(x)$. For $x \in B$, $-(4\pi)^{-1}\Delta\psi(x) = \rho_2^{TF}(x)-\rho_1^{TF}(x) < 0$, so ψ is <u>subharmonic</u> on B (i.e. $\psi(x) \leq$ the average of ψ on any sphere contained in B and centered at x). Hence ψ has its maximum on the boundary of B or at ∞, at all of which points $\psi=0$. Therefore B is the empty set. ∎

In the $\mu=0$ case it is easy to show how Theorem 6 follows from Lemma 8.

<u>Proof of Theorem 6 when $\lambda = \sum_{j=1}^{k} z_j$.</u> The proof when $\lambda < \Sigma z_j$ uses the same ideas but is more complicated. Since $\lambda = \Sigma z_j$ then $\lambda_1 = \sum_{j=1}^{m} z_j$, $\lambda_2 = \sum_{j=m+1}^{k} z_j$ and $\mu=0$ for all three systems. For $\alpha > 0$ let $f(\alpha) = E^{TF}(\alpha z_1,\dots,\alpha z_m, z_{m+1},\dots,z_k; R_1,\dots,R_k) - E^{TF}(\alpha z_1,\dots,\alpha z_m; R_1,\dots,R_m)$ $-E^{TF}(z_{m+1},\dots,z_k; R_{m+1},\dots,R_k)$, where the three E^{TF} are defined for neutral systems (i.e. $\mu=0$ for all α). The goal is to show that $f(1) \geq 0$. Since $f(0) = 0$, it is enough to show that $df(\alpha)/d\alpha \geq 0$. From (26) and (27) it is true, and almost obvious, that

$$\partial E^{TF}/\partial z_i = -\int \rho^{TF}(y)|y-R_i|^{-1}dy + \sum_{j\neq i} z_j|R_i-R_j|^{-1} = \lim_{x\to R_i} \phi^{TF}(x)-z_i|x-R_i|^{-1} .$$

This is the TF version of the Feynman-Hellman theorem; notice how the nuclear-nuclear repulsion comes in here. Thus,

$$df(\alpha)/d\alpha = \sum_{j=1}^{m} \lim_{x\to R_j} \eta_\alpha(x)$$

where $\eta_\alpha(x) = \phi_1^{TF}(x)-\phi_2^{TF}(x)$ and ϕ_1^{TF} is the potential for

$\{\alpha z_1,\dots,\alpha z_m, z_{m+1},\dots,z_k;\ R_1,\dots,R_k\}$ and ϕ_2^{TF} is the potential for $\{\alpha z_1,\dots,\alpha z_m;\ R_1,\dots,R_m\}$. $\phi_1^{TF}(x) \geq \phi_2^{TF}(x)$ for all x by Lemma 8, and hence $\eta_\alpha(x) \geq 0$. ∎

Theorem 6 has a natural application to the stability of matter problem. As will be shown in the next section, the TF energy (27) is, with suitably modified constants, a lower bound to the true quantum energy E_N^Q for <u>all</u> Z. By Theorem 3 (iv) and Theorem 6 we have that

<u>Theorem 9</u>. <u>Fix</u> $\{z_j, R_j\}_{j=1}^k$ <u>and let</u> $Z = \sum_{j=1}^k z_j$. <u>Then for all</u> $\lambda \geq 0$

$$E_\lambda^{TF} \geq E_Z^{TF} \geq -(2.21)q^{2/3}(K^c)^{-1} \sum_{j=1}^k z_j^{7/3} . \tag{43}$$

The latter constant, 2.21, is obtained by numerically solving the TF equation for a single, neutral atom (J. Barnes, private communication). By scaling, (43) holds for any choice of K^c in the definition (26) of $\mathcal{E}(\rho)$.

Theorem 9 is what will be needed for the H-stability of matter, because it says that the TF system is H-stable, i.e. the energy is bounded below by a constant times the nuclear particle number (assuming that the z_j are bounded, of course).

Another application of Theorem 6 that will be needed is the following strange <u>inversion of the role of electrons and nuclei</u> in TF theory. It will enable us to give a lower bound to the true quantum-mechanical <u>electron-electron</u> repulsion. This theorem has nothing to do with quantum mechanics per se; it is really a theorem purely about electrostatics even though it is derived from the TF no binding theorem.

Theorem 10. Suppose that $x_1,\ldots,x_N$ are any N distinct points in 3-space and define

$$V_X(y) = \sum_{j=1}^{N} |y-x_j|^{-1} . \tag{44}$$

Let $\gamma > 0$ and let $\rho(x)$ be any non-negative function such that $\int\rho(x)dx < \infty$ and $\int\rho(x)^{5/3}dx < \infty$. Then

$$\sum_{1\le i<j\le N} |x_i-x_j|^{-1} \ge -\frac{1}{2}\iint\rho(x)|x-y|^{-1}\rho(y)dxdy + \int\rho(y)V_X(y)dy - (2.21)N/\gamma - \gamma\int\rho(y)^{5/3}dy . \tag{45}$$

Proof. Consider $\mathcal{E}(\rho)$ (26) with q=1, k=N, K^c replaced by γ, $z_j \equiv 1$ and $R_j \equiv x_j$, j=1,...,N. Let $\lambda = \int\rho(x)dx$. Then $\mathcal{E}(\rho) \ge E_\lambda^{TF}$ (by definition) and $E_\lambda^{TF} \ge -(2.21)N/\gamma$ by Theorem 9. The difference of the two sides in (45) is just $\mathcal{E}(\rho) + (2.21)N/\gamma$. ∎

IV. The Stability of Bulk Matter

The various results of the last two sections can now be assembled to prove that the ground state energy (or infimum of the spectrum, if this is not an eigenvalue) of H_N is bounded below by an extensive quantity, namely the total number of particles, independent of the nuclear locations $\{R_j\}$. This is called the H-stability of matter to distinguish it from thermodynamic stability introduced in the next section. As explained before, the inclusion of the nuclear kinetic energy, as will be done in the next section, can only raise the energy.

The first proof of the N-boundedness of the energy was given by Dyson and Lenard (Dyson-Lenard, 1967, Lenard-Dyson, 1968). Their proof is a remarkable analytic tour de force, but a chain of sufficiently many inequalities was used that they ended up with an estimate of something like -10^{14} Rydbergs/particle. Using the results of the previous sections we will end up with -23 Rydbergs/particle (see (55)).

We have in mind, of course, that the nuclear charges z_j, if they are not all the same, are bounded above by some fixed charge z.

Take any fermion $\psi(x_1,\ldots,x_N;\sigma_1,\ldots,\sigma_N)$ which is normalized and antisymmetric in the (x_i,σ_i). Define the kinetic energy T_ψ and the single particle density ρ_ψ as in (16) and (17). We wish to compute a lower bound to

$$E_\psi^Q \equiv \langle\psi,H_N\psi\rangle \tag{46}$$

with H_N being the N-particle Hamiltonian given in (23) and $\langle\psi,\psi\rangle = 1$.

For the third term on the right side of (23) Theorem 10 can be used with ρ taken to be ρ_ψ. Then, for any $\gamma > 0$

$$\langle\psi, \sum_{1\le i<j\le N} |x_i - x_j|^{-1}\psi\rangle \ge \frac{1}{2}\iint \rho_\psi(x)|x-y|^{-1}\rho_\psi(y)\,dx\,dy$$

$$-(2.21)N\gamma^{-1} - \gamma\int \rho_\psi(y)^{5/3}dy \qquad (47)$$

Notice how the first and second terms on the right side of (45) combine to give + 1/2 since

$$\langle\psi, \{\int \rho_\psi(y)V_x(y)dy\}\psi\rangle = \iint \rho_\psi(x)|x-y|^{-1}\rho_\psi(y)\,dx\,dy \quad . \qquad (48)$$

To control the kinetic energy in (23) Theorem 1 is used; the total result is then

$$E_\psi^Q \ge \alpha\int \rho_\psi(x)^{5/3}dx - \int V(x)\rho_\psi(x)\,dx + \frac{1}{2}\iint \rho_\psi(x)|x-y|^{-1}\rho_\psi(y)\,dx\,dy$$

$$+ U(\{z_j, R_j\}_{j=1}^k) - (2.21)N\gamma^{-1} \qquad (49)$$

with

$$\alpha = (4\pi q)^{-2/3}\, K^c - \gamma \; . \qquad (50)$$

Restrict γ, which was arbitrary, so that $\alpha > 0$. Then, apart from the constant term $-(2.21)N\gamma^{-1}$, (49) is just $\mathcal{E}_\alpha(\rho_\psi)$, the Thomas-Fermi energy functional $\mathcal{E}$ applied to ρ_ψ, but with $q^{-2/3}K^c$ replaced by α. Since $\mathcal{E}_\alpha(\rho_\psi) \ge E_{\alpha,N}^{TF} \equiv \inf\{\mathcal{E}_\alpha(\rho): \int\rho = N\}$ (by definition), and since the neutral case always has the lowest TF energy, as shown in Theorem 9, we have that

$$\mathcal{E}_\alpha(\rho_\psi) \ge -(2.21)\alpha^{-1} \sum_{j=1}^k z_j^{7/3} \; . \qquad (51)$$

Thus we have proved the following:

<u>Theorem 11.</u> <u>If ψ is a normalized, antisymmetric function of space and spin of N variables, and if there are q spin states associated with each particle then, for any $\gamma > 0$ such that α defined by (50) is positive,</u>

$$\langle\psi, H_N\psi\rangle \geq -(2.21)\{N\gamma^{-1} + \alpha^{-1} \sum_{j=1}^{k} z_j^{7/3}\} . \qquad (52)$$

<u>The optimum choice for γ is</u>

$$\gamma = (4\pi q)^{-2/3} K^c [(\sum_{j=1}^{k} z_j^{7/3}/N)^{1/2} + 1]$$

<u>in which case</u>

$$E_N^Q \geq -(2.21) \frac{(4\pi q)^{2/3}N}{K^c} \{1 + [\sum_{j=1}^{k} z_j^{7/3}/N]^{1/2}\}^2 \quad . \qquad (53)$$

This is the desired result, but some additional remarks are in order.

(1) Since $[1+a^{1/2}]^2 \leq 2+2a$,

$$E_N^Q \geq -(4.42)\ (4\pi q)^{2/3}(K_c)^{-1}\ \{N + \sum_{j=1}^{k} z_j^{7/3}\} . \qquad (54)$$

Thus, provided the nuclear charges z_j are bounded above by some fixed z, E_N^Q is indeed bounded below by a constant times the <u>total</u> particle number N+k.

(2) Theorem 11 does not presuppose neutrality.

(3) For electrons, q=2 and the prefactor in (53) is -(2.08)N. As remarked after Theorem 1, the unwanted constant $(4\pi)^{2/3}$ has been improved to $[4\pi/(1.83)]^{2/3}$. Using this, the prefactor becomes -(1.39)N. If $z_j = 1$ (hydrogen atoms) and N = k (neutrality) then

$$E_N^Q \geq -(5.56)N = -(22.24)N \text{ Ry} . \qquad (55)$$

(4) The power law $z^{7/3}$ cannot be improved upon for large z because Theorem 5 asserts that the energy of an atom is indeed proportional to $z^{7/3}$ for large z.

(5) It is also possible to show that matter is indeed bulky. This will be proved for any ψ and any nuclear configuration (not just the minimum energy configuration) for which $E_\psi^Q \le 0$. The minimizing nuclear configuration is, of course, included in this hypothesis. Then

$$0 \ge E_\psi^Q = \frac{1}{2} T_\psi + \langle\psi, H_N'\psi\rangle$$

where H_N' is (23) but with a factor 1/2 multiplying $\sum_{i=1}^N \Delta_i$. By Theorem (11), $\langle\psi, H_N'\psi\rangle \ge 2E_N$, where E_N is the right side of (53) (replace K^c by $K^c/2$ there). Therefore, the first important fact is that

$$T_\psi \le 4|E_N| \quad ,$$

and this is bounded above by the total particle number.

Next, for any $p \ge 0$, it is easy to check that there is a $C_p > 0$ such that for any nonnegative $\rho(x)$,

$$\{\int\rho(x)^{5/3}dx\}^{p/2} \int|x|^p \rho(x)dx \ge C_p\{\int\rho(x)dx\}^{1+5p/6} \; .$$

It is easy to find a minimizing ρ for this, and to calculate C_p: $\rho(x)^{2/3} = 1-|x|^p$ for $|x| \le 1$; $\rho(x) = 0$, otherwise.

Since T_ψ satisfies (18) we have that

$$\langle\psi, \sum_{i=1}^N |x_i|^p\psi\rangle = \int|x|^p \rho_\psi(x)dx \ge C_p' N(N^{5/3}/|E_N|)^{p/2} \; ,$$

with $C_p' = C_p(K^c/4)^{p/2}(4\pi q)^{-p/3}$.

If it is assumed that $\sum z_j^{7/3}/N$ is bounded, and hence that $(N^{5/3}/|E_N|)^{p/2} > AN^{p/3}$ for some A, we reach the conclusion that the radius of the system is at least of the order $N^{1/3}$, as it should be.

The above analysis did not use any specific property of the Coulomb potential, such as the virial theorem. It is also applicable to the more general Hamiltonian $H_{n,k}$ in (58).

(6) The q dependence was purposely retained in (53) in order to say something about <u>bosons</u>. If q=N, then it is easy to see that the requirement of antisymmetry in ψ is <u>no restriction</u> at all. In this case then, one has simply

$$E_N^Q = \inf \operatorname{spec} H_N$$

over all of $L^2(\mathbb{R}^3)^N$. Therefore

$$E_N^Q(\text{bosons}) \geq \frac{-(2.21)(4\pi)^{2/3}}{K^c} N^{5/3} \{1+[\sum_{j=1}^{k} z_j^{7/3}/N]^{1/2}\}^2 \quad . \qquad (56)$$

It was shown by Dyson and Lenard (Dyson-Lenard, 1967) that

$$E_N^Q(\text{bosons}) \geq -(\text{constant})N^{5/3} \quad ,$$

and by Dyson (Dyson, 1967) that

$$E_N^Q(\text{bosons}) \leq -(\text{constant})\ N^{7/5} \quad . \qquad (57)$$

Proving (57) was not easy. Dyson had to construct a rather complicated variational function related to the type used in the BCS theory of superconductivity. Therefore <u>bosons are not stable</u> under the action of Coulomb forces, but the exact power law is not yet known. Dyson has conjectured that it is 7/5.

In any event, the essential point has been made that Fermi statistics is essential for the stability of matter. The uncertainty principle for one particle, even in the strong form (5), together with intuitive notions that the electrostatic energy ought not to be very great, are insufficient for stability. The additional physical fact that is needed is that the kinetic energy increases as the 5/3 power of the fermion density.

V. The Thermodynamic Limit

Having established that E_N^Q is bounded below by the total particle number, the next question to consider is whether, under appropriate conditions, E_N^Q/N has a limit as $N \to \infty$, as expected. More generally, the same question can be asked about the free energy per particle when the termperature is not zero and the particles are confined to a box.

It should be appreciated that the difficulty in obtaining the lower bound to E_N^Q came almost entirely from the r^{-1} short range singularity of the Coulomb potential. Other potentials, such as the Yukawa potential, with the same singularity would present the same difficulty which would be resolved in the same way. The singularity was tamed by the $\rho^{5/3}$ behavior of the fermion kinetic energy.

The difficulty for the thermodynamic limit is different. It is caused by the long range r^{-1} behavior of the Coulomb potential. In other words, we are faced with the problem of explosion rather than implosion. Normally, a potential that falls off with distance more slowly than $r^{-3-\varepsilon}$ for some $\varepsilon > 0$ does not have a thermodynamic limit. Because the charges have different signs, however, there is hope that a cancellation at large distances may occur.

An additional physical hypothesis will be needed, namely neutrality. To appreciate the importance of neutrality consider the case that the electrons have positive, instead of negative charge. Then $E_N^Q > 0$ because every term in (23) would be positive. While the H-stability question is trivial in this case, the thermodynamic limit is not. If the particles are constrained to be in a domain Ω whose volume $|\Omega|$ is proportional to N, the particles will repel each other so strongly that they will all go to the boundary of Ω in order to minimize the electrostatic energy. The minimum electrostatic energy will be of the order $+N^2|\Omega|^{-1/3} \sim +N^{5/3}$. Hence no thermodynamic limit will exist.

When the system is neutral, however, the energy can be expected to be extensive, i.e. O(N). For this to be so, different parts of the system far from each other must be approximately independent, despite the long range nature of the Coulomb force. The fundamental physical, or rather electrostatic, fact that underlies this is screening; the distribution of the particles must be sufficiently neutral and isotropic locally so that according to Newton's theorem (13 below) the electric potential far away will be zero. The problem is to express this idea in precise mathematical form.

We begin by defining the Hamiltonian for the entire system consisting of k nuclei, each of charge z and mass M, and n electrons ($\hbar^2/2 = 1$, $m = 1$, $|e| = 1$):

$$H_{n,k} = -\sum_{j=1}^{n} \Delta_j - \frac{1}{M}\sum_{j=n+1}^{n+k} \Delta_j - z\sum_{i=1}^{n}\sum_{j=n+1}^{n+k} |x_i - y_j|^{-1}$$

$$+ \sum_{1\le i<j\le n} |x_i - x_j|^{-1} + z^2 \sum_{n+1\le i<j\le n+k} |y_i - y_j|^{-1} . \qquad (58)$$

The first and second terms in (58) are, respectively, the kinetic energies of the electrons and the nuclei. The last three terms are, respectively, the electron-nuclear, electron-electron and nuclear-nuclear Coulomb interactions. The electron coordinates are x_i and the nuclear coordinates are y_i. The electrons are fermions with spin 1/2; the nuclei may be either bosons or fermions.

The basic neutrality hypotheses is that n and k are related by

$$n = kz . \qquad (59)$$

It is assumed that z is rational.

The thermodynamic limit to be discussed here can be proved under more general assumptions, i.e. we can have several kinds of negative particles (but they must all be fermions in order that the basic stability estimate of Section IV holds) and several kinds of nuclei with different statistics, charges and masses. Neutrality must always hold, however. Short range forces and hard cores, in addition to the Coulomb forces, can also be included with a considerable sacrifice in simplicity of the proof.

$H_{n,k}$ acts on square integrable functions of n+k variables (and spin as well). To complete the definition of $H_{n,k}$ we must specify boundary conditions: choose a domain Ω (an open set, which need not be connected) and require that $\psi = 0$ if x_i or y_i are on the boundary of Ω.

For each non-negative integer j, choose an n_j and a corresponding k_j determined by (59), and choose a domain Ω_j. The symbol N_j will henceforth stand for the pair (n_j,k_j) and

$$|N_j| \equiv n_j+k_j \quad .$$

We require that the densities

$$\rho_j \equiv |N_j| \; |\Omega_j|^{-1} \tag{60}$$

be such that

$$\lim_{j\to\infty} \rho_j = \rho \quad . \tag{61}$$

ρ is then the density in the thermodynamic limit. Here we shall choose the Ω_j to be a sequence of balls of radii R_j and shall denote them by B_j.

It can be shown that the same thermodynamic limit for the energy and free energy holds for any sequence N_j,Ω_j and depends only on the limiting

ρ and β, and not on the "shape" of the Ω_j, provided the Ω_j go to infinity in some reasonable way.

The basic quantity of interest is the <u>canonical partition function</u>

$$Z(N,\Omega,\beta) = \mathrm{Tr}\exp(-\beta H_{n,k}) \tag{62}$$

where the trace is on $L^2(\Omega)^{|N|}$ and $\beta = 1/T$, T being the temperature in units in which Boltzmann's constant is unity.

The <u>free energy per unit volume</u> is

$$F(N,\Omega,\beta) = -\beta^{-1}\ln Z(N,\Omega,\beta)/|\Omega| \tag{63}$$

and the problem is to show that with

$$F_j = F(N_j,\Omega_j,\beta) \tag{64}$$

then

$$\lim_{j\to\infty} F_j \equiv F(\rho,\beta) \tag{65}$$

exists. A similar problem is to show that

$$E(N,\Omega) \equiv |\Omega|^{-1}\inf_{\psi} \langle\psi,H_{n,k}\psi\rangle/\langle\psi,\psi\rangle, \tag{66}$$

the <u>ground state energy per unit volume</u>, has a limit

$$e(\rho) = \lim_{j\to\infty} E_j \tag{67}$$

where

$$E_j = E(N_j,\Omega_j) \quad .$$

The proof we will give for the limit $F(\rho,\beta)$ will hold equally well for $e(\rho)$ because E_j can be substituted for F_j in all statements.

The basic strategy consists of two parts. The easiest part is to show that F_j is bounded below. We already know this for E_j by the results of section IV. The second step is to show that in some sense the sequence F_j is decreasing. This will then imply the existence of a limit.

Theorem 12. Given N,Ω and β there exists a constant C depending only on $\rho = |N|/|\Omega|$ and β such that

$$F(N,\Omega,\beta) \geq C \quad . \tag{68}$$

Proof. Write $H = H_A + H_B$, where

$$H_A = -\frac{1}{2}\{\sum_{i=1}^{n}\Delta_i + \frac{1}{M}\sum_{j=n+1}^{n+k}\Delta_j\}$$

is half the kinetic energy. Then $H_B \geq b|N|$, with b depending only on z, by the results of Section IV (increasing the mass by a factor of 2 in H_B only changes the constant b). Hence $Z(N,\Omega,\beta) \leq e^{-\beta b|N|}\mathrm{Tr}\exp(-\beta H_A)$. However, $\mathrm{Tr}\exp(-\beta H_A)$ is the partition function of an ideal gas and it is known by explicit computation that it is bounded above by $e^{\beta d|N|}$ with d depending only on $\rho = |N|/|\Omega|$ and β. Thus

$$F(N,\Omega,\beta) \geq (b-d)\rho \quad . \quad ▮$$

For the second step, two elementary but basic inequalities used in the general theory of the thermodynamic limit are needed and they will be described next.

A. Domain partition inequality: Given the domain Ω and the particle numbers $N = (n,k)$, let π be a partition of Ω into ℓ disjoint domains $\Omega^1,\ldots,\Omega^\ell$. Likewise N is partitioned into ℓ integral parts (some of which may be zero):

$$N = N^1+\ldots+N^\ell \quad .$$

Then for any such partition, π, of Ω and N

$$Z(N,\Omega,\beta) = \mathrm{Tr}\exp(-\beta H_{n,k}) \geq \mathrm{Tr}^{\pi}\exp(-\beta H_N^{\pi}) \tag{69}$$

Here Tr^{π} means trace over

$$\mathcal{H}^{\pi} \equiv L^2(\Omega^1)^{|N^1|} \otimes \ldots \otimes L^2(\Omega^{\ell})^{|N^{\ell}|} \quad ,$$

and H_N^{π} is defined as in (58) but with Dirichlet ($\psi = 0$) boundary conditions for the N^i particles on the boundary of Ω^i (for $i=1,\ldots,\ell$).

Simply stated, the first N^1 particles are confined to Ω^1, the second N^2 to Ω^2, etc. The interaction among the particles in different domains is still present in H_N^{π}. (69) can be proved by the Peierls-Bogoliubov variational principle for Tr e^X. Alternatively, (69) can be viewed simply as the statement that the insertion of a hard wall, infinite potential on the boundaries of the Ω^i only decreases Z; the further restriction of a definite particle number to each Ω^i further reduces Z because it means that the trace is then over only the H_N^{π}-invariant subspace, $\mathcal{H}^{\pi}$, of the full Hilbert space.

B. <u>Inequality for the interdomain interaction</u>: The second inequality is another consequence of the convexity of $A \to \mathrm{Tr}\, e^A$ (Peierls-Bogoliubov inequality):

$$\mathrm{Tr}\, e^{A+B} \geq \mathrm{Tr}\, e^A \exp \langle B \rangle \tag{70}$$

where

$$\langle B \rangle \equiv \mathrm{Tr}\, B e^A / \mathrm{Tr}\, e^A \quad . \tag{71}$$

Some technical conditions are needed here, but (70) and (71) will hold in our application.

To exploit (70), first make the same partition π as in inequality A and then write

$$H_N^\pi = H_o + W(X) \tag{72}$$

$$H_o = H^1 + \ldots + H^\ell \tag{73}$$

with H^i being that part of the total Hamiltonian (58) involving only the N^i particles in Ω^i, and H^i is defined with the stated Dirichlet boundary conditions on the boundary of Ω^i. $W(X)$, with X standing for all the coordinates, is the inter-domain Coulomb interaction. In other words, $W(X)$ is that part of the last three terms on the right side of (58) which involves coordinates in different blocks of the partition π. Technically, W is a small perturbation of H_o.

With

$$A = -\beta H_o \text{ and } B = -\beta W \tag{74}$$

in (70), we must calculate $\langle W\rangle$. Since $e^A = e^{-\beta H_o}$ is a simple tensor product of operators on each $L^2(\Omega^i)^{|N^i|}$, W is merely the <u>average</u> inter-domain Coulomb energy in a canonical ensemble in which the Coulomb interaction is present in each subdomain but the ℓ domains are independent of each other. This basic idea is due to Griffiths (Griffiths, 1969). In other words, let $q^i(x)$, $x \in \Omega^i$, denote the average <u>charge</u> density in Ω^i for this ensemble of independent domains, namely

$$q^i(x) = \sum_{j=1}^{|N^i|} q_j \int_{|\Omega^i|^{|N^i|-1}} \exp(-\beta H^i)(X^i,X^i)\, \widehat{dx_j}/Z(N^i,\Omega^i,\beta) \tag{75}$$

with the notation: X^i stands for the coordinates of the $|N^i|$ particles in Ω^i, $\widehat{dx_j}$ means integration over all these coordinates (in Ω^i) with the exception of x_j, and x_j is set equal to x; q_j is the charge (-1 or +z) of the jth particle; $\exp(-\beta H^i)(X^i,Y^i)$ is a kernel (x-space representation) for $\exp(-\beta H^i)$. $q^i(x)$ vanishes if $x \notin \Omega^i$.

With the definitions (75) one has that

$$\langle W\rangle = \sum_{i<j} \int_{\Omega_i} \int_{\Omega_j} q^i(x)q^j(y)\,|x-y|^{-1}\,dxdy \quad . \tag{76}$$

(70), together with (76) and (74) is the desired inequality for the inter-domain interaction. It is quite general in that an analogous inequality holds for arbitrary two-body potentials. Neither specific properties of the Coulomb potential nor neutrality was used.

Now we come to the crucial point at which screening is brought in. The following venerable result from the Principia Mathematica is essential.

Theorem 13 (Newton). Let $\rho(x)$ be an integrable function on 3-space such that $\rho(x) = \rho(y)$ if $|x| = |y|$ (isotropy) and $\rho(x) = 0$ if $|x| > R$ for some $R > 0$. Let

$$\phi(x) = \int \rho(y)\,|x-y|^{-1}\,dy \tag{77}$$

be the Coulomb potential generated by ρ. Then if $|x| \geq R$

$$\phi(x) = |x|^{-1} \int \rho(y)\,dy \quad . \tag{78}$$

The important point is that an isotropic, neutral charge distribution generates zero potential outside its support, irrespective of how the charge is distributed radially.

Suppose that N^i is neutral, i.e. the electron number = z times the nucleon number for each subdomain in Ω. Suppose also that the subdomain Ω^i is a ball of radius R^i centered at a^i. Then since H^i is rotation invariant, $q^i(x) = q^i(y)$ if $|x-a^i| = |y-a^i|$, $\int q^i(x)dx = 0$ (by neutrality) and $q^i(x) = 0$ if $|x-a^i| > R^i$. Then, by Theorem 13, every term in (76) involving q^i vanishes, because when $j\neq i$, $q^j(y) = 0$ if $|y-a^i| < R^i$ since

Ω^j is disjoint from Ω^i. Consequently the average interdomain interaction, $<W>$, vanishes.

In the decomposition, π, of Ω into $\Omega^1,\ldots,\Omega^\ell$ and N into $N^1,\ldots,N^\ell$ we will arrange matters such that

(i) $\Omega^1,\ldots,\Omega^{\ell-1}$ are balls

(ii) $N^1,\ldots,N^{\ell-1}$ are neutral

(iii) $N^\ell = 0$.

Then $<W> = 0$ and, using (69) and (70)

$$Z(N,\Omega,\beta) \geq \mathrm{Tr}^\pi \exp(-\beta H_N^\pi) \geq \prod_{i=1}^{\ell-1} Z(N^i,\Omega^i,\beta)e^{-\beta<W>}$$

$$= \prod_{i=1}^{\ell-1} Z(N^i,\Omega^i,\beta) \quad . \tag{79}$$

In addition to (i), (ii), (iii) it will also be necessary to arrange matters such that when Ω is a ball B_K in the chosen sequence of domains, then the sub-domains $\Omega^1,\ldots,\Omega^{\ell-1}$ in the partition of B_K are also smaller balls in the same sequence. With these requirements in mind the standard sequence, which depends on the limiting density, ρ, is defined as follows:

(1) Choose $\rho > 0$.

(2) Choose any N_0 satisfying the neutrality condition (59).

(3) Choose R_0 such that

$$28(4\pi/3)\rho R_0^3 = |N_0| \ . \tag{80}$$

(4) For $j \geq 1$ let

$$R_j = (28)^j R_0 ,$$

$$N_j = (28)^{3j-1} N_0 \tag{81}$$

be the radius of the ball B_j and the particle number in that ball.

It will be noted that the density in all the balls except the first is

$$\rho_j = \rho, \quad j \geq 1 , \tag{82}$$

while the density in the smallest ball is much bigger:

$$\rho_0 = 28\rho \quad . \tag{83}$$

This has been done so that when a ball B_K, $K \geq 1$ is packed with smaller balls in the manner to be described below, the density in each ball will come out right; the higher density in B_0 compensates for the portion of B_K not covered by smaller balls. The radii increase geometrically, namely by a factor of 28.

The number 28 may be surprising until it is realized that the objective is to be able to pack B_K with balls of type B_{K-1}, B_{K-2}, etc. in such a way that as much as possible of B_K is covered and <u>also</u> that very little of B_K is covered by very small balls. If the ratio of radii were too close to unity then the packing of B_K would be inefficient from this point of view. In short, if the number 28 is replaced by a much smaller number the analogue of the following basic geometric theorem will not be true.

<u>Theorem 14 (Cheese theorem)</u>. <u>For j a positive integer define the integer $m_j \equiv (27)^{j-1}(28)^{2j}$. Then for each positive integer $K \geq 1$ it is possible to pack the ball B_K of radius R_K (given by 81) with $\bigcup_{j=0}^{K-1}$ (m_{K-j} balls of radius R_j). "Pack" means that all the balls in the union are disjoint.</u>

We will not give a proof of Theorem 14 here, but note that it entails showing that m_1 balls of radius R_{K-1} can be packed in B_K in a cubic array, then that m_2 balls of radius R_{K-2} can be packed in a cubic array in the interstitial region, etc.

Theorem 14 states that B_K can be packed with $(28)^2$ balls of type B_{K-1}, $(27)(28)^3$ balls of type B_{K-2}, etc. If f_{K-j} is the fraction of the volume of B_K occupied by all the balls of radius R_j in the packing, then

$$f_j = m_j(R_{K-j}/R_K)^3 = \frac{1}{27}\gamma^j \tag{84}$$

with

$$\gamma = 27/28 < 1 \ . \tag{85}$$

The packing is <u>asymptotically complete</u> in the sense that

$$\lim_{K\to\infty} \sum_{j=0}^{K-1} f_{K-j} = (1/27) \sum_{j=1}^{\infty} \gamma^j = 1 \ . \tag{86}$$

It is also "geometrically rapid" because the fraction of $|B_K|$ that is uncovered is

$$\sum_{j=K+1}^{\infty} f_j = \gamma^K \ . \tag{87}$$

The necessary ingredients having been assembled, we can now prove

<u>Theorem 15</u>. <u>Given</u> ρ <u>and</u> $\beta > 0$, <u>the thermodynamic limits</u> $F(\rho,\beta)$ <u>and</u> $e(\rho)$ (65,67) <u>exist for the sequence of balls and particle numbers specified by</u> (80) <u>and</u> (81).

<u>Proof</u>. Let F_K given by (64) be the free energy per unit volume for the ball B_K with N_K particles in it. For $K \geq 1$, partition B_K into disjoint domains $\Omega^1,\ldots,\Omega^\ell$, where the Ω^i for $i=1,\ldots,\ell-1$ designate the smaller balls referred to in Theorem 14, and Ω^ℓ (which is the "cheese" after the holes have been removed) is the remainder of B_K. The smaller balls are copies of B_j, $0 \leq j \leq K-1$; in each of these place N_j particles according to (81). $N^\ell = 0$. The total particle number in B_K is then

$$\sum_{j=0}^{K-1} N_j m_{K-j} =$$

$$N_0\{(27)^{K-1}(28)^{2K} + \sum_{j=1}^{K-1} (28)^{3j-1}(27)^{K-j-1}(28)^{2K-2j}\} = N_0(28)^{3K-1} = N_K$$

as it should be.

Use the basic inequality (79); $\langle W \rangle = 0$ since all the smaller balls are neutral and Ω^ℓ contains no particles. Thus, taking logarithms and dividing by $|B_K|$, we have for $K \geq 1$ that

$$F_K \leq \sum_{j=0}^{K-1} F_j \, f_{K-j} \tag{88}$$

with $f_j = \gamma^j/27$ and $\gamma = 27/28$. This inequality can be rewritten as

$$F_K = \sum_{j=0}^{K-1} F_j \, \gamma^{K-j}/27 - d_K \tag{89}$$

with $d_K \geq 0$. (89) is a <u>renewal equation</u> which can be solved explicitly by inspection:

$$F_K = -\gamma \, d_K - \sum_{j=1}^{K} d_j/28 + F_0/28 \quad . \tag{90}$$

We now use the first step, Theorem 13, on the boundedness of F_K. Since $F_K \geq C$, $\sum_{j=1}^{\infty} d_j$ must be finite, for otherwise (90) would say that $F_K \to -\infty$. The convergence of the sum implies that $d_K \to 0$ as $K \to \infty$. Hence the limit exists; specifically

$$F = \lim_{K\to\infty} F_K = -\sum_{j=1}^{\infty} d_j/28 + F_0/28 \quad . \quad ▮ \tag{91}$$

Theorem 15 is the desired goal, namely the existence of the thermodynamic limit for the free energy (or ground state energy) per unit volume. There are, however, some additional points that deserve comment.

A. For each given limiting density ρ, a particular sequence of domains, namely balls, and particle numbers was used. It can be shown that the same limit is reached for general domains, with some mild conditions on their shape including, of course, balls of different radii than that used here. The argument involves packing the given domains with balls of the standard sequence and vice versa. The proof is tedious, but standard, and can be found in (Lieb-Lebowitz, 1972).

B. Here we have considered the thermodynamic limit for real matter, in which all the particles are mobile. There are, however, other models of some physical interest. One is Jellium in which the positive nuclei are replaced by a fixed, uniform background of positive charge. With the aid of an additional trick the thermodynamic limit can also be proved for this model (Lieb-Narnhofer, 1975). Another, more important model is one in which the nuclei are fixed point charges arranged periodically in a lattice. This is the model of solid state physics. Unfortunately, local rotation invariance is lost and Newton's Theorem 13 cannot be used. This problem is still open and its solution will require a deeper insight into screening.

C. An absolute physical requirement for $\beta F(\rho,\beta)$, as a function of $\beta = 1/T$, is that it be concave. This is equivalent to the fact that the specific heat is non-negative since (specific heat) $= -\beta^2 \partial^2 \beta F(\rho,\beta)/\partial\beta^2$. Fortunately it is true. From the definitions (57), (58) we see that $\ln Z(N,\Omega,\beta)$ is convex in β for every finite system and hence $\beta F(N,\Omega,\beta)$ is concave. Since the limit of a sequence of concave functions is always concave, the limit $\beta F(\rho,\beta)$ is concave in β.

D. Another absolute requirement is that $F(\rho,\beta)$ be convex as a function of ρ. This is called thermodynamic stability as distinct from the lower bound H-stability of the previous sections. It is equivalent to the fact that the compressibility is non-negative, since $(\text{compressibility})^{-1} = \partial P/\partial\rho = \rho\, \partial^2 F(\rho,\beta)/\partial\rho^2$. Frequently, in approximate theories (e.g. Van der Waals' theory of the vapor-liquid transition, some field theories, or some theories of magnetic systems in which the magnetization per unit volume plays the role of ρ), one introduces an F with a double bump. Such an F is non-physical and never should arise in an exact theory.

For a finite system, F is defined only for integral N, and hence not for all real ρ. It can be defined for all ρ by linear interpolation, for example, but even so it can neither be expected, nor is it generally, convex, except in the limit. The idea behind the following proof is standard.

Theorem 16. The limit function $F(\rho,\beta)$ is a convex function of ρ for each fixed β. $E(\rho)$ is also a convex function of ρ.

Proof: This means that for $\rho = \lambda\rho^1+(1-\lambda)\rho^2$, $0 \le \lambda \le 1$,

$$F(\rho,\beta) \le \lambda F(\rho^1,\beta)+(1-\lambda)F(\rho^2,\beta) \tag{92}$$

and similarly for $E(\rho)$. As F is bounded above on bounded ρ intervals (this can be proved by a simple variational calculation), it is sufficient to prove (92) when $\lambda = 1/2$. To avoid technicalities (which can be supplied) and concentrate on the main idea, we shall here prove (92) when ρ^2 and ρ^1 are rationally related: $a\rho^1 = b\rho^2$, a, b, positive integers. Choose any neutral particle number M and define a sequence of balls B_j

with radii as given in (81) and with $28(4\pi/3)\rho R_0^3 = (a+b)|M|$. For the ρ system take $N_0 = (a+b)M$, $N_j = (28)^{3j-1} N_0$, $j \geq 1$. For the ρ^1 (resp. ρ^2) system take $N_0^1 = 2b\ M$, $N_j^1 = (28)^{3j-1} N_0^1$ (resp. $N_0^2 = 2aM$, $N_j^2 = (28)^{3j-1} N_0^2$). Consider the ρ system. In the canonical partition, π, of B_K into smaller balls (Theorem 14) note that the number of balls B_j is m_{K-j} and this number is <u>even</u>. In <u>half</u> of these balls place N_j^1 particles and in the other half place N_j^2 particles, $0 \leq j \leq K-1$. Then in place of (88) we get

$$F_K(\rho) \leq \frac{1}{2} \sum_{j=0}^{K-1} f_{K-j} \, [F_j(\rho^1) + F_j(\rho^2)] \tag{93}$$

in an obvious notation. Inserting (89) on the right side of (93),

$$F_K(\rho) \leq \frac{1}{2} [F_K(\rho^1) + F_K(\rho^2)] + \frac{1}{2} (d_K^1 + d_K^2) \, . \tag{94}$$

Since $\lim_{K\to\infty} d_K^{1,2} = 0$, we can take the limit $K \to \infty$ in (94) and obtain (92). ∎

<u>E.</u> The convexity in ρ^1 and concavity in β of $F(\rho,\beta)$ has another important consequence. Since F is bounded below (Theorem 13) and bounded above (by a simple variational argument) on bounded sets in the (ρ,β) plane, the convexity/concavity implies that it is <u>jointly continuous</u> in (ρ,β). This, together with the monotonicity in K of $F_K + \gamma\ d_K$ (see (90)), implies by a standard argument using Dini's theorem that <u>the thermodynamic limit is uniform</u> on bounded (ρ,β) sets. This uniformity is sometimes overlooked as a basic desideratum of the thermodynamic limit. Without it one would have to fix ρ and β <u>precisely</u> in taking the limit — an impossible task experimentally. With it, it is sufficient to have merely an increasing sequence of systems such that $\rho_j \to \rho$ and $\beta_j \to \beta$. The same result holds for $e(\rho)$.

F. An application of the uniformity of the limit for $e(\rho)$ is the following. Instead of confining the particles to a box (Dirichlet boundary condition for $H_{n,k}$) one could consider $H_{n,k}$ defined on all of $L^2(\mathbb{R}^3)^{|N|}$, i.e. no confinement at all. In this case

$$E_N^Q \equiv \inf_\psi \langle\psi, H_{n,k}\psi\rangle / \langle\psi,\psi\rangle$$

is just the ground state energy of a neutral molecule and it is expected that $E_N^Q/|N|$ has a limit. Indeed, this limit exists and it is simply

$$\lim_{n\to\infty} E_N^Q/|N| = \lim_{\rho\to 0} \rho^{-1} e(\rho) \quad .$$

There is no analogue of this for $F(\rho,\beta)$ because removing the box would cause the partition function to be infinite even for a finite system.

G. The ensemble used here is the canonical ensemble. It is possible to define and prove the existence of the thermodynamic limit for the microcanonical and grand canonical ensembles and to show that all three ensembles are equivalent (i.e. that they yield the same values for all thermodynamic quantities, such as the pressure). (See Lieb-Lebowitz, 1972).

H. Charge neutrality was essential for taming the long range Coulomb force. What happens if the system is not neutral? To answer this let N_j, Ω_j be a sequence of pairs of particle numbers and domains, but without (59) being satisfied. Let $Q_j = zk_j - n_j$ be the net charge, $\rho_j = |N_j|/|\Omega_j|$ as before, and $\rho_j \to \rho$. One expects that if

(i) $Q_j|\Omega_j|^{-2/3} \to 0$ then the same limit $F(\rho,\beta)$ is achieved as if $Q_j = 0$. On the other hand, if

(ii) $Q_j|\Omega_j|^{-2/3} \to \infty$ then there is no limit for $F(N_j,\Omega_j,\beta)$. More precisely $F(N_j,\Omega_j,\beta) \to \infty$ because the minimum electrostatic energy is too great. Both of these expectations can be proved to be correct.

The interesting case is if

(iii) $\lim_{j\to\infty} Q_j|\Omega_j|^{-2/3} = \sigma$ exists. Then one expects a <u>shape dependent limit</u> to exist as follows. Assume that the Ω_j are geometrically similar, i.e. $\Omega_j = \lambda\Omega_0$ with $|\Omega_0| = 1$ and $|N_j|\lambda^{-3} = \rho_j$ with $\rho_j \to \rho$. Let C be the <u>electrostatic capacity</u> of Ω_0; it depends upon the shape of Ω_0. The capacity of Ω_j is then $C_j = C\,\lambda$. From elementary electrostatics theory the expectation is that

$$\lim_{j\to\infty} F(N_j,\Omega_j,\beta) = F(\rho,\beta)+\sigma^2/2C\ . \qquad (95)$$

Note that $(Q_j^2/2C_j)|\Omega_j|^{-1} \to \sigma^2/2C$.

<u>(95) can be proved</u> for ellipsoids and balls. The proof is as complicated as the result is simple. With work, the proof could probably be pushed through for other domains Ω_0 with smooth boundaries.

The result (95) is amazing and shows how special the Coulomb force is. It says that the surplus charge Q_j goes to a thin layer near the surface. There, only its electrostatic energy, which overwhelms its kinetic energy, is significant. The bulk of Ω_j is neutral and uninfluenced by the surface layer because the latter generates a constant potential inside the bulk. It is seldom that one has two strongly interacting subsystems and that the final result has no cross terms, as in (95).

<u>I.</u> There might be a temptation, which should be avoided, to suppose that the thermodynamic limit describes a single phase system of uniform

density. The temptation arises from the construction in the proof of Theorem 15 in which a large domain B_K is partitioned into smaller domains having essentially constant density. Several phases can be present inside a large domain. Indeed, if β is very large a solid is expected to form, and if the average density, ρ, is smaller than the equilibrium density, ρ_s, of the solid a dilute gas phase will also be present. The location of the solid inside the larger domain will be indeterminate.

From this point of view, there is an amusing, although expected, aspect to the theorem given in (95). Suppose that β is very large and that $\rho < \rho_s$. Suppose, also, that a surplus charge $Q = \sigma V^{2/3}$ is present, where V is the volume of the container. In <u>equilibrium</u>, the surplus charge will never be bound to the surface of the solid, for that would give rise to a larger free energy than in (95).

As a final remark, the existence of the thermodynamic limit (and hence the existence of intensive thermodynamic variables such as the pressure) does not establish the existence of a <u>thermodynamic state</u>. In other words, it has not been shown that correlation functions, which always exist for finite systems, have limits as the volume goes to infinity. Indeed, unique limits might not exist if several phases are present. For well behaved potentials there are techniques available for proving that a state exists when the density is small, but these techniques do not work for the long-range Coulomb potential. Probably the next chapter to be written in this subject will consist of a proof that correlation functions are well defined in the thermodynamic limit.

VI. Hartree-Fock Theory

As a practical matter, a good estimate for E_N^Q (see (23) and (25)), even with fixed nuclei, is difficult to obtain. An old method (Hartree, 1927, Fock, 1930, Slater, 1930) is still much employed. Indeed, chemists refer to it as an ab initio calculation.

Without taking any position on the usefulness of a HF calculation, it might be worthwhile to present the results of recent work (Lieb-Simon 1973) to the effect that HF theory is at least well defined, i.e. the HF equations have solutions. Unlike the situation for Thomas-Fermi theory, the solutions are not unique in general.

To define HF theory let

$$\psi = \{\phi_1,\dots,\phi_N\} \tag{96}$$

denote a set of N single particle functions of space and spin, ϕ_i, in $L^2(\mathbb{R}^3;\mathbb{C}^2)$. Two spin states are assumed here. Form the Slater determinant

$$D_\psi(x_1,\dots,x_N;\sigma_1,\dots,\sigma_N) \equiv (N!)^{-1/2} \det|\phi_i(x_j,\sigma_j)|_{i,j=1}^N \tag{97}$$

which is an antisymmetric function of space-spin. The $L^2(\mathbb{R}^3;\mathbb{C}^2)^N$ norm of D_ψ is

$$\langle D_\psi,D_\psi\rangle = \det|M_{ij}^\psi|_{i,j=1}^N \tag{98}$$

where M^ψ is the overlap (Gram) matrix:

$$M_{ij}^\psi = \langle\phi_i,\phi_j\rangle = \sum_{\sigma=1}^{2}\int \overline{\phi_i}(x,\sigma)\phi_j(x,\sigma)\,dx\,. \tag{99}$$

The HF energy is

$$E_N^{HF} = \inf\{\langle D_\psi,H_N D_\psi\rangle : \langle D_\psi,D_\psi\rangle = 1\} \tag{100}$$

and by the variational principle

$$E_N^{HF} \geq E_N^Q\,. \tag{101}$$

When the ϕ_i are orthonormal one easily finds that

$$<D_\psi, H_N D_\psi> \equiv \mathcal{E}_N(\psi) + U(\{z_j, R_j\}_{j=1}^k) \tag{102}$$

where

$$\mathcal{E}_N(\psi) = \sum_{i=1}^N <\nabla\phi_i, \nabla\phi_i> + <\phi_i, [V + \tfrac{1}{2} W_\psi - \tfrac{1}{2} K_\psi]\phi_i> . \tag{103}$$

V is given in (24a). W_ψ is a multiplication operator depending on $x \in \mathbb{R}^3$ and not on σ:

$$W_\psi(x) = \sum_{i=1}^N \sum_{\sigma=1}^2 \int |\phi_i(y,\sigma)|^2 \, |x-y|^{-1} \, dy . \tag{104}$$

K_ψ is a more complicated, spin dependent integral operator:

$$(K_\psi \phi)(x,\sigma) = \sum_{i=1}^N \phi_i(x,\sigma) \int (\phi_i|\phi)(y) \, |x-y|^{-1} \, dy \tag{105}$$

where

$$(f|g)(y) = \sum_{\sigma=1}^2 \overline{f(y,\sigma)} \, g(y,\sigma) \tag{106}$$

is a partial inner product. It is important to note that W_ψ and K_ψ in (103) do not depend on i. In the earlier Hartree theory, which will not be discussed here, the analogue of (103) does contain i dependent operators. Nevertheless, the theorems to be presented here do carry over, <u>mutatis mutandis</u>, to Hartree theory.

E_N^{HF} is finite. The basic question is whether there is a minimizing ψ for E_N^{HF}. This is, of course, the same as minimizing $\mathcal{E}_N(\psi)$. What should one expect? If $N < Z+1$ (the "less than" is important) there should be a minimum because otherwise there would be an uncompensated positive Coulomb potential at large distances which, however weak, can always bind an additional electron. (Recall that $Z = \Sigma z_j$ and $z_j > 0$; Z need not be an integer.) If $N \geq Z+1$ the existence of a minimum is less obvious. It may or may not occur, depending on the details of the nuclear configuration. We shall have nothing to say about this latter case.

Theorem 17. If $N < Z+1 = 1+\Sigma z_j$ then, for any nuclear configuration, there is a minimizing ψ for E_N^{HF}. Furthermore, the ϕ_i in ψ can be chosen to be orthonormal, i.e. $M_{ij}^{\psi} = \delta_{ij}$.

The proof of Theorem 17 involves a trick which in retrospect is obvious, but which took some time to notice. Here is an outline

(1) Consider $\mathcal{E}_N(\psi)$ as defined by (103), (104) and (105). This is a quartic expression in the ϕ_i. Both $\mathcal{E}_N(\psi)$ and $\langle D_\psi, D_\psi \rangle$ are invariant under any unitary transformation of the form

$$\phi_i \to \sum_{j=1}^{N} R_{ji}\, \phi_j \tag{107}$$

with R being an N × N unitary matrix. If R is chosen to diagonalize M^{ψ}, we can restrict our attention to ψ such that the ϕ_i are orthogonal.

The minimizing ψ will be constructed by taking a weak limit of a sequence $\psi^{(n)}$ such that

$$U + \lim_{n\to\infty} \mathcal{E}_N(\psi^{(n)}) = E_N^{HF} \quad . \tag{108}$$

The major difficulty is that a weak limit of orthonormal functions need not be orthogonal. It could even happen that $\lim_{n\to\infty} \phi_i^{(n)} = \phi$ (independent of i). The trick to overcome the difficulty is this: Instead of minimizing $\mathcal{E}_N(\psi)$ subject to $\langle D_\psi, D_\psi \rangle = 1$ consider instead

$$e_N = \inf\{\mathcal{E}_N(\psi) : \psi \in S_N\} \tag{109}$$

with

$$S_N = \{\psi : M^{\psi} \leq I_N\} \ . \tag{110}$$

I_N is the N × N identity matrix and the inequality in (110) is that $I_N - M^\psi$ is positive semidefinite. The obvious, but crucial, fact is that a weak limit of functions in S_N remains in S_N.

If there is a minimizing ψ for e_N, the ϕ_i can be chosen to be orthogonal, possibly after a unitary transformation (107). Then, since $\psi \in S_N$, $\langle\phi_i,\phi_i\rangle = \delta_i \leq 1$. Assume $\delta_i > 0$, all i. Then to see that the δ_i can be chosen to be unity, note that $\mathcal{E}_N(\psi)$ is quadratic in each ϕ_i. Therefore if ϕ_i is replaced by $(\gamma_i/\delta_i)^{1/2}\,\phi_i$, with $\gamma_i > 0$, $\mathcal{E}_N$ is linear in each γ_i. Clearly $\partial\mathcal{E}_N/\partial\gamma_i \leq 0$ (otherwise $\mathcal{E}_N$ can be decreased in taking $\gamma_i=0$, which contradicts the assumption that $\langle\phi_i,\phi_i\rangle > 0$ at the minimum), and thus $\mathcal{E}_N$ is not increased if γ_i is taken to be 1.

The problem, then, is to show two things:

(2a) there is a minimizing ψ for e_N;

(2b) M^ψ does not have a zero eigenvalue.

(2a). This is an application of functional analysis. Given (108) one can find a sequence $\psi^{(n)}$ such that each $\phi_i^{(n)}$ converges weakly to ϕ_i in the Sobolev space $W^1(\mathbb{R}^3)$, i.e.

$$\phi_i^{(n)} \overset{w}{\to} \phi_i \quad\text{and}\quad \nabla\phi_i^{(n)} \overset{w}{\to} \nabla\phi_i \ . \tag{111}$$

The functional $\sum_{i=1}^{N}\langle\phi_i,[W_\psi-K_\psi]\phi_i\rangle$ is weakly lower semicontinuous, essentially because $W_\psi-K_\psi$ is a positive operator and is bounded on $W^1(\mathbb{R}^3)$. The positivity of the function $|x-y|^{-1}$ on $\mathbb{R}^6$ is used. Finally,

$$\langle\phi_i^{(n)},V\phi_i^{(n)}\rangle \to \langle\phi_i,V\phi_i\rangle$$

because V is a relatively compact perturbation of $-\Delta$ in the sense of quadratic forms. Thus ψ minimizes $\mathcal{E}_N(\psi)$ on S_N.

(2b). If M^ψ has a zero eigenvalue then $e_N = e_{N-1}$, i.e. one of the ϕ_i vanishes (after a unitary transformation (107)). This is impossible if $N < Z+1$ because one can always find a ϕ orthogonal to $\phi_1,\ldots,\phi_{N-1}$ such that $\mathcal{E}_N(\phi_1,\ldots,\phi_{N-1},\phi) < \mathcal{E}_{N-1}(\phi_1,\ldots,\phi_{N-1})$. The property of the hydrogenic Hamiltonian (1) that it has infinitely many negative eigenvalues is used in an essential way.

By a standard argument in the calculus of variations, the minimizing ψ satisfies the Euler-Lagrange equation for $\mathcal{E}_N(\psi)$ as follows.

<u>Theorem 18. Let</u> $\psi = (\phi_1,\ldots,\phi_N)$ <u>be any minimizing</u> ψ <u>for</u> E_N^{HF} <u>arranged such that</u> $M^\psi = I_N$. <u>It is not necessary to assume that</u> $N < Z+1$. <u>Let</u> H_ψ <u>be the operator on</u> $L^2(\mathbb{R}^3,\mathbb{C}^2)$:

$$H_\psi = -\Delta + V + W_\psi - K_\psi \tag{112}$$

<u>as defined in (104) and (105). Then for</u> $i = 1,\ldots,N$

(i)
$$H_\psi\phi_i = \lambda_i\phi_i \tag{113}$$

<u>for some</u> $\lambda_i < 0$. <u>This is the HF equation.</u>

(ii) $\{\lambda_1,\ldots,\lambda_N\}$ <u>are the lowest N eigenvalues of</u> H_ψ.

It is not true that $e_N = \sum_{i=1}^{N} \lambda_i$ because there are no factors of 1/2 in (112). The only slightly unusual point is (ii) which follows from the fact that $\mathcal{E}_N(\psi)$ is quadratic in each ϕ_i. If some eigenvalue of H_ψ among the lowest N is missing, e_N can be lowered by using the missing eigenfunction instead of the (N+j)th eigenfunction.

In summary, just as in the analogous case of TF theory, it has been shown that the nonlinear HF equation (113) not only has a solution, but that among these solutions there is one that minimizes the HF energy E_N^{HF}. It is not easy to prove directly that (113) has solutions.

In general, it is difficult to say much about a minimizing ψ. Despite the deceptive notation, (113) is not a linear equation. One cannot say, as one could for the linear Schroedinger equation, that the ϕ_i can be assumed to be real or that $\phi_i(x,\sigma)$ is a product $f_i(x)g_i(\sigma)$. These assumptions are often made in practice. What can be done is to restrict the ϕ_i from the beginning to be real and/or product functions such that for any $i \neq j$ $g_i = g_j$ or g_i is othogonal to g_j. Then the whole analysis can be done afresh and Theorems 17 and 18 will hold. The minimum, $\tilde{E}_N^{HF}$, in this restricted class might be greater than E_N^{HF}, however. In the same manner other restrictions can be placed on the ϕ_i (such as rotation invariance, for example) with the same conclusion. The only essential requirement is that for any ϕ and $\psi = \{\phi_i\}$ in the restricted class, $H_\psi \phi$ is in the same class.

The overriding question is, of course, how close is E_N^{HF} to E_N^Q? It is difficult to give a precise answer, but in two limiting cases HF theory is exact. One is the hydrogen atom; the other is the $Z \to \infty$ limit. It was in fact a determinantal wave function (97), not the best one to be sure, that was used in the variational upper bound leading to Theorem 5. Thus

$$\lim_{Z\to\infty} Z^{-7/3} E_N^{HF} = \lim_{Z\to\infty} Z^{-7/3} E_N^Q \tag{114}$$

in the sense of Theorem 5.

References

1. Balàzs, N., 1967, Formation of stable molecules within the statistical theory of atoms, Phys. Rev. **156**, 42-47.

2. Birman, M.S., 1961, Mat. Sb. **55** (97), 125-174; The spectrum of singular boundary value problems, Amer. Math. Soc. Transl. Ser. 2 (1966), **53**, 23-80.

3. Dirac, P.A.M., 1930, Note on exchange phenomena in the Thomas atom, Proc. Camb. Phil. Soc. **26**, 376-385.

4. Dyson, F.J., 1967, Ground-state energy of a finite system of charged particles, J. Math. Phys. **8**, 1538-1545.

5. Dyson, F.J. and A. Lenard, 1967, Stability of matter. I, J. Math. Phys. **8**, 423-434.

6. Fermi, E., 1927, Un metodo statistico per la determinazione di alcune prioretà dell' atome, Rend. Acad. Naz. Lincei **6**, 602-607.

7. Fock, V., 1930, Näherungsmethode zur Lösung des quantenmechanischen Mehrkörperproblems, Zeit. Phys. **61**, 126-148; see also V. Fock, "Selfconsistent field" mit austausch für Natrium, Zeit. Phys. **62** (1930), 795-805.

8. Gombás, P., 1949, "Die statistischen Theorie des Atomes und ihre Anwendungen", Springer Verlag, Berlin.

9. Griffiths, R.B., 1969, Free energy of interacting magnetic dipoles, Phys. Rev. **172**, 655-659.

10. Hartree, D.R., 1927-28, The wave mechanics of an atom with a non-Coulomb central field. Part I. Theory and methods, Proc. Camb. Phil. Soc. **24**, 89-110.

11. Heisenberg, W., 1927, Über den anschaulichen Inhalt der quantentheoretischen Kinematik und Mechanik, Zeits. Phys., 43, 172-198.

12. Jeans, J.H., 1915, The mathematical theory of electricity and magnetism, Cambridge University Press, third edition, page 168.

13. Kirzhnits, D.A., 1957, J. Exptl. Theoret. Phys. (U.S.S.R.) 32, 115-123. Engl. transl. Quantum corrections to the Thomas-Fermi equation, Sov. Phys. JETP, 5 (1957), 64-71.

14. Kompaneets, A.S. and E.S. Pavlovskii, 1956, J. Exptl. Theoret. Phys. (U.S.S.R.) 31, 427-438. Engl. transl. The self-consistent field equations in an atom, Sov. Phys. JETP, 4 (1957), 328-336.

15. Lenard, A. and F.J. Dyson, 1968, Stability of matter. II, J. Math. Phys. 9, 698-711.

16. Lenz, W., 1932, Über die Anwendbarkeit der statistischen Methode auf Ionengitter, Zeit. Phys. 77, 713-721.

17. Lieb, E.H., 1976, Bounds on the eigenvalues of the Laplace and Schroedinger operators, Bull. Amer. Math. Soc., in press.

18. Lieb, E.H. and J.L. Lebowitz, 1972, The constitution of matter: existence of thermodynamics for systems composed of electrons and nuclei, Adv. in Math. 9, 316-398. See also J.L. Lebowitz, and E.H. Lieb, Existence of thermodynamics for real matter with Coulomb forces, Phys. Rev. Lett. 22 (1969), 631-634.

19. Lieb, E.H. and H. Narnhofer, 1975, The thermodynamic limit for jellium, J. Stat. Phys. 12, 291-310. Erratum: J. Stat. Phys. 14 (1976), No. 5.

20. Lieb, E. H. and B. Simon, 1973, On solutions to the Hartree-Fock problem for atoms and molecules, J. Chem. Phys. 61, 735-736. Also a longer paper in preparation.

21. Lieb, E.H. and B. Simon, 1975, The Thomas-Fermi theory of atoms, molecules and solids, Adv. in Math., in press. See also E.H. Lieb and B. Simon, Thomas-Fermi theory revisited, Phys. Rev. Lett. **33** (1973), 681-683.

22. Lieb, E.H. and W.E. Thirring, 1975, A bound for the kinetic energy of fermions which proves the stability of matter, Phys. Rev. Lett. **35**, 687-689, Errata: Phys. Rev. Lett. (1975), **35**, 1116. For more details on kinetic energy inequalities and their application, see also E.H. Lieb and W.E. Thirring, Inequalities for the moments of the Eigenvalues of the Schrödinger Hamiltonian and their relation to Sobolev inequalities, in Studies in Mathematical Physics: Essays in Honor of Valentine Bargmann, E.H. Lieb, B. Simon and A.S. Wightman editors, Princeton University Press, 1976.

23. Rosen, G., 1971, Minimum value for c in the Sobolev inequality $\|\phi\|_3 \leq c\|\nabla\phi\|^3$, SIAM J. Appl. Math. **21**, 30-32.

24. Schwinger, J., 1961, On the bound states of a given potential, Proc. Nat. Acad. Sci. (U.S.) **47**, 122-129.

25. Scott, J.M.C., 1952, The binding energy of the Thomas Fermi atom, Phil. Mag. **43**, 859-867.

26. Sheldon, J.W., 1955, Use of the statistical field approximation in molecular physics, Phys. Rev. **99**, 1291-1301.

27. Slater, J.C., 1930, The theory of complex spectra, Phys. Rev. **34**, 1293-1322.

28. Sobolev, S.L., 1938, Mat. Sb. **46**, 471 (1938). See also S.L. Sobolev, Applications of functional analysis in mathematical physics, Leningrad (1950), Amer. Math. Soc. Transl. of Monographs, **7** (1963).

29. Sommerfeld, A., 1932, Asymptotische Integration der Differentialgleichung des Thomas-Fermischen Atoms, Zeit. Phys. 78, 283-308.

30. Teller, E., 1962, On the stability of molecules in the Thomas-Fermi theory, Rev. Mod. Phys. 34, 627-631.

31. Thomas, L.H., 1927, The calculation of atomic fields, Proc. Camb. Phil. Soc. 23, 542-548.

32. Von Weizsäcker, C.F., 1935, Zur Theorie der Kernmassen, 96, 431-458.

CENTRO INTERNAZIONALE MATEMATICO ESTIVO

(C.I.M.E.)

REPORT ON RENORMALIZATION GROUP

B. TIROZZI

Istituto di Matematica, Università di Camerino

Corso tenuto a Bressanone dal 21 giugno al 24 giugno 1976

REPORT ON RENORMALIZATION GROUP

Prof. Benedetto Tirozzi

Istituto di Matematica

Università di Camerino

Introduction

1. Integral and local central limit theorems of probability theory and the renormalization group method.

In this lecture we want to present the problem of the research of automodel probability distributions in comparison with usual integral and local central limit theorems. We think that this approach is instructive for understanding the main mathematical idea underlyng this kind of problems.

Consider a stationary discrete random field ξ_j, $j \in Z^1$ and the sequence $S_K = \sum_{j=1}^{K} \xi_j$ and suppose that $E\xi_j = 0$. Then the random field $\{\xi_j\}$ satisfies the integral central limit theorem if

1.(1) $$\lim_{K\to\infty} P_e\left\{ \frac{S_K}{\sqrt{DS_K}} < z \right\} \longrightarrow \frac{1}{\sqrt{2\pi}} \int_{-\infty}^{z} e^{-\frac{u^2}{2}} du$$

This theorem was proven in the pioneer works of Gnedenko (1),(2),(3) and Kolmogorov in the case in which ξ_j are independent and equally distributed. For us it is more interesting the case in which the random variables ξ_i, are not independent and more precisely when $\{\xi_i, i \in Z^1\}$ form a Gibbs random field, (4),

(5), corresponding to a certain potential

1. 2)
$$U = \begin{cases} +\beta\,\Phi(\xi_t, t\in K) & |K|>1,\ K\in Z^1 \\ -\mu\xi_t & |K|=1 \end{cases}$$

where we suppose that $\xi_t \in X$, X being the space of realization of the Gibbs random field.

It is well known, from very general arguments, that a necessary and sufficient condition for 1.1 to be true is that the field $\xi_t, t\in Z^1$ must be strong mixing (2) with mixing coefficient $\alpha(n)$ and that the dispersion $D\xi_k \sim CK$ where C is some positive constant, further it is required a condition of the type of Lindeberg (1),(2), analogous to the one used for the case of independent variables:

1. 3)
$$\frac{n}{pDS_n} \int_{|z|\geq\varepsilon\sqrt{DS_n}} z^2\, dF_p(z) \xrightarrow[n\to\infty]{} 0$$

where $F_p(z) = P_r\{\xi_1 + \dots + \xi_p < z\}$ is some positive constant and $p = o(n)$.

The coefficient of strong mixing is defined in the following way: let $(\Omega, \mathcal{F}, P)$ be a probability space where the random variables ξ_i are defined and let us call $\mathfrak{M}_a^b$ the minimal σ-algebra generated by the events

$$A = \{(\xi_{t_1}, \dots, \xi_{t_s}) \in \bar{A}\}$$

where $\bar{A} \subset X^s$ and $a \leq t_1 \leq t_2 \dots \leq t_s \leq b$.

Then we can define the coefficient of mixing by the following quantity':

1. 4)
$$\alpha(K) = \sup_{\substack{A\in\mathfrak{M}_{-\infty}^0 \\ B\in\mathfrak{M}_K^\infty}} |P(A\cap B) - P(A)\,P(B)|$$

and we will say that the field $(\xi_t, t\in Z^1)$ is strong mixing if we have that

1. 5)
$$\alpha(K) \xrightarrow[K\to\infty]{} 0$$

Now we can show how it is possible from the knowledge of 1. 5 to deduce the behaviour of DS_K .

In fact from a very general theorem (2) for stationary processes we have that

1. 6) $$|E\xi_t\xi_s| \le c\,\alpha(|s-t|) \qquad c>0$$

And so we can give the following estimate for the dispersion of S_n supposing $E\xi_t=0$

1. 7) $$DS_n \le \sum_{t,s\in(1,\dots,n)} |E\xi_t\xi_s| \le \sum_{t,s\in(1,\dots,n)} \alpha(|s-t|).$$

Suppose now that $\sum_{t\in Z^1}\alpha(|s-t|)<B$ for all s then we obtain suddenly

1. 8) $$DS_n \sim B\cdot n\ , \quad B>0$$

that is the required condition on DS_n

The proof of the central limit theorem under the hypothesis said above proceeds in the following way. Divide the interval $[1,n]$ in segments V_i, W_i such that

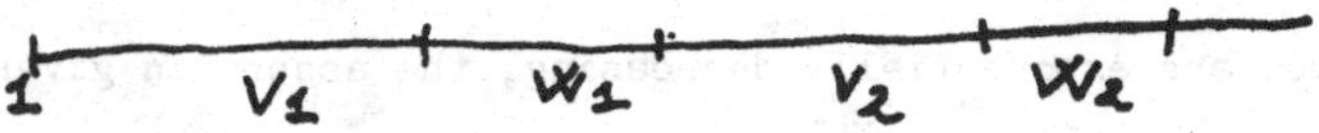

fig.1.

$|V_i|=p,\ |W_i|=q=o(p)$ for all i and so we have that

1. 9) $$\lim_{n\to\infty}\frac{|W_i|}{|V_i|}=0\ , \qquad \lim_{n\to\infty}\frac{|V_i|}{n}=0$$

It can be shown that the random variables $S_{V_i}=\frac{1}{n^{1/2}}\sum_{t\in V_i}\xi_t$ are asymptotically independent if

1. 10) $$\lim_{n\to\infty}\left[\frac{n}{p+q}\right]\alpha(q)=0$$

In this case we obtain the following asymptotic joint probability distribution for the variables $S_{V_1},\dots,S_{V_j},\dots$

1. 11) $$P_r^{(n)}\{S_{V_1}\le z_1,\dots,S_{V_n}\le z_n\}\xrightarrow[n\to\infty]{}\int_{-\infty}^{z_1}\frac{e^{-u_1^2/2}}{\sqrt{2\pi}}du_1\cdots\int_{-\infty}^{z_n}\frac{e^{-u_n^2/2}}{\sqrt{2\pi}}du_n$$

while using 1.8 it can be seen that the variables

1. 12) $$S_{W_i} = \frac{1}{n^{\frac{1}{2}}} \sum_{t \in W_i} \xi_t$$

go in probability to zero.

The idea of this method (method of deletions) is due to Bernstein (2).

It is possible to find in the papers on Statistical Physics or on Probability theory many cases in which the central integral limit theorem has been derived for Gibbs fields with general potentials given by 1.2 using the very fundamental ideas here exposed or procedures which are founded on them.

For example Nakhapitan (6) has generalized the method of Bernstein's deletions used in (2) for a Gibbs field in any dimension and there he used the condition of the uniform strong mixing for the Gibbs field which he showed to be true for a certain class of potentials. Malishev (7) has been able to show that, if the correlations defined in 1.6 of the Gibbs field, when the space of realization is finite, are exponentially decreasing, the assertion given in 1.11 is true: he uses the technique of evaluating semiinvariantes. He is also able to show that integral central limit theorem holds for every translationally invariant states in the case of Ising models for $\beta > \beta_{cr}$ and $\beta < \beta_{cr}$. Then from a theorem shown by R. L. Dobrushin and B. Tirozzi (8) it follows that also the local limit theorem is true for these Gibbs states. These methods together with other founded on properties of analyticity (9), (10) show that the integral and local central limit theorem are true for $\beta \gg \beta_{cr}$ and $\beta \ll \beta_{cr}$ for a large class of system of interest of Statistical Physics.

Now we want to begin to examine the situation when $\beta = \beta_{cr}$. The first observation consists of the fact that it is no more true that

1. 13) $$D S_n \sim c n$$

because of the smaller rate of decreasing of correlations and so we expect a normalization factor given by $n^{\frac{\alpha}{2}}, 1 < \alpha < 2$. Furthermore, given a certain potential we are interested not only in the existence of the asymptotic joint probability distributions for the normed sums of spins but also in the determination of β_{cr} for a particular hamiltonian and also it is important to know how stable is β_{cr} with respect to "small" changes of the potential.

As an example of the situation of the known results about these problems we finish this introduction showing an open problem for a onedimensional spin system with a pair potential given by

$$V(s,t) = \frac{1}{|s-t|^{\alpha}} \qquad s,t \in Z^1$$

when $\beta << \beta_{cr}$.

The question is the following: is the local central limit theorem verified in this case ? We know $E(\xi_{t_1}\xi_{t_2}) \sim \frac{\text{const}}{|t_1-t_2|^{\alpha}}$ but we have no informations about the coefficient of mixing.

2. Kadanoff renormalization group and gaussian automodel distributions.

We are going to give a more precise formulation of the above problems and to describe some examples of limit probability distributions.

Let R be the space of all real numbers. We shall consider a random field $\xi_t, t \in Z^1$ which takes values in R.

Definition 2.1 Let us consider a realization $\xi = \{\xi_i\}_{-\infty}^{+\infty}$ of the random field, $\xi \in R^{Z^1}$. We will define an endomorphism on R^{Z^1} in such a way:

2.1.
$$\begin{cases} A_K(\alpha) : \xi \to \tilde{\xi} & K \in Z^1 \\ \tilde{\xi}_s = (A_K(\alpha)\xi)_s = \dfrac{1}{K^{\alpha/2}} \displaystyle\sum_{e=sK}^{K(s+1)-1} \xi_e & 1<\alpha<2 \end{cases}$$

The transformation $A_K^*(\alpha)$ adjoint of $A_K(\alpha)$ acting in the space of probability measures defined on R^{Z^1} is defined by

2.2.
$$A_K^*(\alpha)P(C) = \tilde{P}(C) = P(A_K^{-1}C)$$

where $C \in \mathcal{B}_R^{Z^1}$ i.e. C belongs to the $\sigma-$ algebra generated by the random field $\xi_t, t \in Z^1$.

Definition 2.2. A probability distribution on R^{Z^1} is a automodel if

$$A_K^*(\alpha)P = \tilde{P} = P \tag{2.3.}$$

or, in other words, P is a automodel if it is a fixed point of the transformation $A_K^*(\alpha)$.

Remark 1. $A_K(\alpha)$ is an endomorphism of R^{Z^1} in R^{Z^1} and was first introduced by Kadanoff (11). We describe explicity, for sake of clearness, the transformation 2.1.

Consider the one dimensional lattice Z^1 and suppose that to each point i of the lattice is associated the value taken by the random variable ξ_i

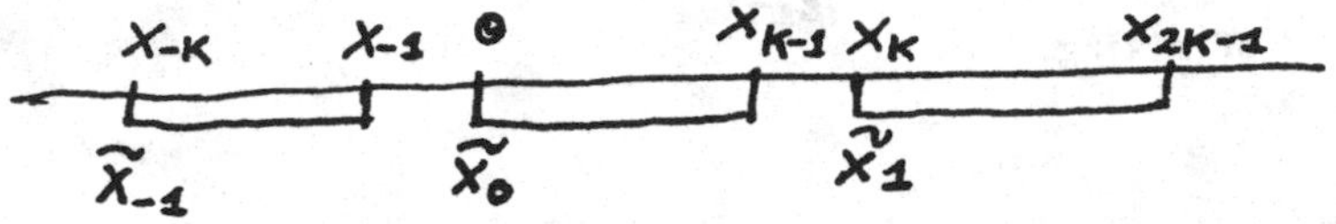

fig. 2.

Divide the lattice in blocks of length K as in fig. 2., then sum all the variables X_i which are in the same block, multiply for the normalization factor $K^{-\frac{\alpha}{2}}$ give a numeration to the blocks in the following way: the block having an extreme left spin $X_{\ell K}$ has index ℓ .

Then the transformated sequence will be

$$\tilde{X}_{-1}, \tilde{X}_0, \tilde{X}_1, \tilde{X}_2, \dots$$

where $\tilde{X}_i$ is equal to the weighted sum of the variables in the block i defined above. We can find the probability measure $\tilde{P}'$ defined on R^{Z^1} using definition 2.2. Let P be given by

$$P(C) = \int_{\xi \in C} P(d(\xi)), \text{ where } \xi \in R^{Z^1}, \ \xi = \{\xi_i\}_{-\infty}^{+\infty} \tag{2.4.}$$

$C \in \mathcal{F}_{R^{Z^1}}$ then $\tilde{P}'$ is the probability measure defined on R^{Z^1} generated by the following measure on cylindrical sets B_n :

2.5. $$\tilde{P}(B_n) = \int_{A_K\xi \in B_n} P(d\xi)$$

where $$B_n = \{ \xi_{-n} = \tilde{x}_{-n}, \ldots, \xi_n = \tilde{x}_n \}$$

Remark 2. It is easy to ses that the family of operators $\{A_K(\alpha)\}$ forms a multiplication semi-group of endomorphisms, that is $\forall K, q \in (Z^1)^+ A_q(\alpha)A_K(\alpha) = A_{Kq}(\alpha)$. From this simple observation it follows easily the equivalence between definition 2.2 and the formulation given in the preceeding paragraph.

In fact the search of a fixed point for $A_K(\alpha)$ is related to the consideration of the limit

$$\lim_n (A_K^*(\alpha))^n P = \lim_n (A_{K^n}^*(\alpha)) P$$

and this corresponds to the investigation of the asymptotic joint probability distribution of the random variables

2.6. $$\frac{1}{(K^n)^{\alpha/2}} \sum_{e=j(K^n)}^{K^n(j+1)-1} \xi_e \quad , j = 1, 2,$$

We are able to determine the spectral density of an automodel stationary field $\{\xi_i\}_{-\infty}^{+\infty}$. Let us introduce the generalized random function $\eta(t) = \sum_{n=-\infty}^{+\infty} \xi_n e^{int}$, $\eta(t)$ is the Fourier transform of the field $\{\xi_n\}_{-\infty}^{+\infty}$ and has the following properties

2.7. $$\eta(t+2\pi) = \eta(t) \quad , \overline{\eta(t)} = \eta(-t)$$

We look for the transformation induced on it by Kadanoff's renorm group:

2.8. $$\xi'_1 = \frac{1}{K^{\alpha/2}} \sum_{j=1K}^{(1+1)K-1} \xi_j$$

Calling $\eta'(t)=\sum_{n=-\infty}^{+\infty} e^{int}\xi'_n$ the generalized random function obtained Fourier transforming the new field $\{\xi'_n\}_{-\infty}^{+\infty}$ we have that $A_K(\alpha)$ wi induce a transformation on $\eta(t)$: $Q_K(\alpha)\,\eta(t)\rightarrow\eta'(t)$.

We shall find it with the help of very simple calculations.

In fact we have

2.9.

$$\eta'(t)=\sum_{n=-\infty}^{+\infty}\frac{e^{int}}{K^{\alpha/2}}\sum_{j=Kn}^{K(n+1)-1}\xi_j=\sum_{n=-\infty}^{+\infty}\frac{e^{int}}{K^{\alpha/2}}\sum_{q=Kn}^{K(n+1)-1}\frac{1}{2\pi}\int_{-\pi}^{+\pi}e^{-iqt'}\eta(t')dt'$$

$$=\frac{K^{-\frac{\alpha}{2}}}{2\pi}\int_{-\pi}^{+\pi}\sum_{n=-\infty}^{+\infty}e^{-iKnt'}e^{int}\frac{e^{-it'K}-1}{e^{-it'}-1}\eta(t')dt'=$$

$$=K^{-\frac{\alpha}{2}}\int_{-\pi}^{+\pi}\delta_p(t-Kt')\frac{e^{-it'K}-1}{e^{-it'}-1}\eta(t')dt'=$$

$$=K^{-\frac{\alpha}{2}-1}\int_{-\pi K}^{+\pi K}\delta_p(t-Kt')\frac{e^{-i\tau}-1}{e^{-i\frac{\tau}{K}}-1}\eta\left(\frac{\tau}{K}\right)d\tau=$$

$$=K^{-1-\frac{\alpha}{2}}\int_{-\pi}^{+\pi}\delta_p(t-\tau)\sum_{j=1}^{K}\frac{e^{-i\tau}-1}{e^{-i\left(\frac{\tau}{K}+\frac{2\pi}{K}j\right)}-1}\eta\left(\frac{\tau}{K}+\frac{2\pi}{K}j\right)d\tau=$$

$$=K^{-1-\frac{\alpha}{2}}\sum_{j=1}^{K}\left(e^{-it}-1\right)\frac{\eta\left(\frac{t}{K}+\frac{2\pi}{K}j\right)}{e^{-i\left(\frac{t}{K}+\frac{2\pi}{K}j\right)}-1}$$

If we set $\theta(t)=(e^{-it}-1)^{-1}\eta(t)$ we obtain the following equation of transformation:

2.10.
$$\theta'(t)=K^{-1-\frac{\alpha}{2}}\sum_{j=0}^{K-1}\theta\left(\frac{t}{K}+\frac{2\pi}{K}j\right)$$

Since the field $\{\xi_i\}_{-\infty}^{+\infty}$ is stationary we have that the correlation function of the generalized process $\eta(t)$ can be expressed in terms of the spectral density of the field $\{\xi_i\}_{-\infty}^{+\infty}$.

This follows from elementary calculations.

2.11. $$E(\eta(t)\eta(t')) = \sum_{n,m} e^{int+imt'} E(\xi_n \xi_m) =$$
$$= \sum_{n} \sum_{m-n} E(\xi_0 \xi_{m-n}) e^{i(m-n)t'+in(t+t')} =$$
$$= \delta_p(t+t')\rho(t)$$

where
$$\rho(t) = \sum_p E(\xi_0 \xi_p) e^{ipt}$$

We have also that

2.12. $$(e^{-it}-1)(e^{-it'}-1) E(\theta(t)\theta(t')) = E(\eta(t)\eta(t'))$$

From 2.10 we obtain the law of transformation for $\psi(t)$ the correlation function of $\theta(t)$

2.13. $$\psi'(t) = K^{-1-\alpha} \sum_{j=1}^{K} \psi\left(\frac{t}{K} + \frac{2\pi}{K} j\right)$$

Let us look for its fixed points. We choose as input function

2.14. $$\psi^{(0)}(t) = |t|^{-\gamma}$$

We will indicate the transformation 2.13 with the symbol $R_K(\alpha)$, the n we have to calculate the following limit keeping in mind remark 2):

2.15. $$\lim_{K\to\infty} R_K(\alpha)\,\psi^{(0)}$$

We have

2.16. $$R_K(\alpha)\psi^{(0)} = K^{-1-\alpha} \sum_{j:-K\le \frac{t}{2\pi}+j\le K} \left|\frac{t}{K} + \frac{2\pi}{K} j\right|^{-\gamma} = K^{\gamma-1-\alpha} \sum |t+2\pi j|^{-\gamma}$$

We have a limit only in the case $\gamma=1+\alpha$ and so we obtain the explicit expression of

$$2.17. \quad \psi(t) = \sum_{j=-\infty}^{+\infty} \frac{1}{|t+2\pi j|^{1+\alpha}}$$

The spectral function of the process $\{\xi_i\}_{-\infty}^{+\infty}$ can be obtained by using 2.11 and 2.12:

$$2.18. \quad \rho(t) = |e^{-it}-1|^2 \sum_{j=-\infty}^{+\infty} \frac{1}{|t+2\pi j|^{1+\alpha}}$$

We have the following condition on α: $1<\alpha<2$ the first limit is due to the consideration made in the introduction, the second is due also to the fact that we want the divergence in the origin to be integrable.
So we obtain that the function $\rho(t)$ has in the origin a singolarity of the type $|t|^{-\alpha+1}$

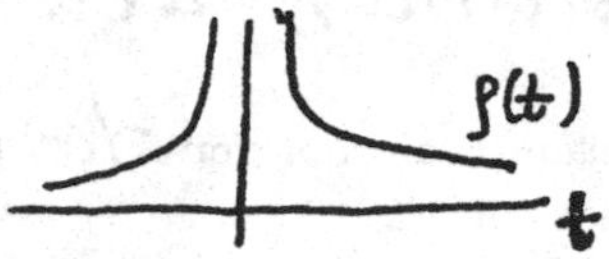

and the correlation function of the process will have the following behaviour at infinity $b(\tau) \sim |\tau|^{-(1-\alpha)-1} = |\tau|^{\alpha-2}$

3. <u>Stability of the automodell probability distributions, non gaussian automodel distribution</u> *)

In the last section we have shown how it is possible to find gaussian automodel probability distributions: in fact we have calculated the correlation function of the automodel field so if it is gaussian, it is completely determined once we have fixed the condition $E\,\xi_1=0$

Now we want to show that apparently there exist also others automodel pr.

---

*) In this section are exposed the principal concepts contained in the paper (12).

distr., i.e. non gaussian a.p.d. and we are also going to give a method for finding these new types of a.pr.d.

We will take a point of view in analogy of the classical bifurcation theory of dynamical systems (13), (14), (15).

Let us consider a gaussian a.p.d. $\mathcal{G}_\alpha$ of the Kadanoff's renormalization group $A_K(\alpha)$ in the one dimensional case. We will define the "differential" of the renormalization group $\partial_K A^*(\alpha)$ acting in the "tangent space" to the probability distributions: then we study the spectrum of $\partial_K A^*(\alpha)$ and we look for those values of α, $1<\alpha<2$ for which there appears an eigenvalue of $\partial_K A^*(\alpha)$ equal to one.

In this case we expect that in the neighbourhood of the corresponding α the gaussian a.p.distr. $\mathcal{G}_\alpha$ looses its stability and that near this $\mathcal{G}_\alpha$ there appears a new branch of automodel non gaussian distributions.

Before going into the details of this construction let us define more exactly what we mean under stability in our case.

Let us take as an example the classical integral central limit theorem of Probability theory. Let $A_2(1)$ be the Kadanoff's renormalization group obtained for $K=2$, $\alpha=1$ then we consider the transformation $A_2^*(1)$ acting on probability densities $p(x)$

3.1. $$\left(A_2^*(1)p\right)(x)=\int_{-\infty}^{+\infty} p(\sqrt{2}x-y)\,p(y)\,dy$$

It is evident that the limit distribution obtained by applying $(A_2(1))^n$, $n\to\infty$ to the field $\{\xi_i\}_{-\infty}^{+\infty}$ must coincide with the limit distribution of

3.2. $$S_K=\frac{\xi_1+\cdots+\xi_{2^K}}{\sqrt{2^K}}$$

and, if $\{\xi_i\}_{-\infty}^{+\infty}$ is a field of independent random variables equally distributed with $E\xi_i=0$ then the dispersion, $D\xi_i=\sigma^2$ coincide with the dispersion of the limit law:

3.3. $e^{-x^2/\sigma^2}/\sqrt{2\pi}\sigma$

the "fixed point" of $A_2^*(1)$ will satisfy the following equation:

$$p(x) = \int p(\sqrt{2}x-y)\,p(y)\,dy$$

So we can state that the "manifold" of the initial random fields, the point of which will converge to the probability distribution 3.2 when $A_2(1)$ is applied n times, $n\to\infty$, is determined by the dispersion.

In the case of $\alpha>1$ this argument does not hold and we will take the idea of stability from the classical bifurcation theory.

Let $T: M\to M$ be a diffeomorfism of a n-dimensional manifold (16) in itself and let $x\in M$ be a fixed point of T. (fig. 3). Let $\gamma(\beta)$ be a curve in M parametrized by β and let $\tilde{M}$ be a n-l submanifold of M such that if $x_0\in\tilde{M}$ then $T^n x_0 \xrightarrow[n\to\infty]{} x$ if $M = M'\oplus\tilde{M}$ and $\gamma(\beta)$ belongs to M' then there is only one intersection between $\gamma(\beta)$ and M which will be obtained for a certain value of β.

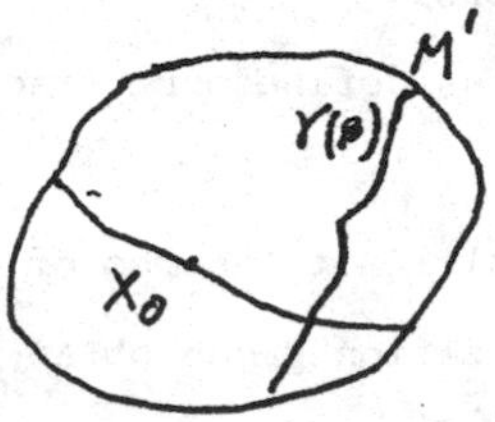

fig. 3

This situation is achieved when, and only when, the differential of T in X has only one eigenvalue bigger than one, then M' is a direction such that T restricted on M' is expanding and T restricted on $\tilde{M}$ is contracting.

In the case when M is the "manifold" of probability distributions, T is the renormalization group $A_K^*(\alpha)$, x is P_α that is a a.p.d. We have that $\tilde{M}$ is a "manifold" of probability distributions such that $A_K^*(\alpha)$ acting on it is contracting and M' is the set of initial pr.distr. and clearly the value of β shown before will be the critical temperature. If such a situation is verified we shall say that P_α is stable, thus P_α will loose its stability when there will be two eigenvalues bigger or equal than one and then we expect the appearing of a new branch of automodel probability distributions.

Now we will enter more into the details and give an explicit construction of the tangent space to a gaussian aut.prob.distr. We will write formal expressions for sake of semplicity but it is possible to give to them an exact and rigorous meaning **using the same procedure as in (17).**

Let G_α be a gaussian stationary aut.pr.distr. on R^{Z^1} and let $b(\tau)$:

3.4. $$b(\tau) = E(\xi_\tau \xi_0) = \int_{-\frac{1}{2}}^{+\frac{1}{2}} e^{2\pi i \lambda \tau} \rho(\lambda) d\lambda$$

be the correlation function, where $\rho(t)$ is given by 2,18. Define the matrix $B = \|b_{k\ell}\|$ setting $b_{k,\ell} = b(k-\ell)$ and $A = \|a_{k\ell}\| = \|b_{k\ell}\|^{-1}$ where

$$a_{k,\ell} = a_{k-\ell} = \int_{-\frac{1}{2}}^{+\frac{1}{2}} e^{2\pi i (k-\ell)\lambda} \rho^{-1}(\lambda) d\lambda$$

Then G_α can be written formally as

3.5. $$G_\alpha = e^{-\sum_{i,j \in Z^1} a_{i-j} \xi_i \xi_j}$$

For defining the tangent space in G_α let us consider the set of stationary probability distributions "near" to G_α in the sense that they are absolutely continuous with respect to it

3.6. $$G_\alpha^\varepsilon = e^{-\sum_{i,j \in Z^1} a_{i-j} \xi_i \xi_j - \varepsilon \sum_{i,j,k,\ell \in Z^1} a_{i,j,k,\ell} \xi_i \xi_j \xi_k \xi_\ell}$$

where $a_{i,j,k,\ell}$ are real numbers such that: $a_{i+m,j+m,k+m,\ell+m} = a_{i,j,k,\ell}$

The density of G_α^ε with respect to G_α will be considered as a vector of the tangent space, so we will give the following definitions.

<u>Definition</u> 3.1. A form of degree n of the process $\{\xi_i\}_{-\infty}^{+\infty}$ will be the random variable given by the following formal sum

3.7. $$a^{(n)} = \sum_{\ell_1,\dots,\ell_n \in Z^1} a_{\ell_1 \dots \ell_n} \xi_{\ell_1} \cdots \xi_{\ell_n}$$

where $a_{\ell_1 \dots \ell_n}$ are real numbers and symmetric, in $\ell_1, \dots, \ell_n$.

The forms of degree form a vector space which we will indicate $\mathcal{Y}_n$. We can introduce in $\mathcal{Y}_n$ the topology given by the convergence of the coordinates.

Definition 3.2. A stationary form of degree n will be a form $a^{(n)} \in \mathcal{Y}_n$ such that $a_{e_1+m,\dots,e_n+m}$ does not depend on $m \in Z^1$

The space of stationary forms of degree n will be indicated as $\mathcal{Y}_n^{(st)}$

It is evident that $\mathcal{Y}_n^{(st)}$ is a closed subspace of $\mathcal{Y}_n$

A large class of stationary forms of degree n can be obtained in the following way. Let $\alpha(\lambda_1,\dots,\lambda_n)$ be a continuous symmetric function defined on the n dimensional torus Tor n such that its Fourier series is absolutely convergent for $\lambda_j \in \mathrm{Tor}_1$. Let us consider the form

3.8. $$a^{(n)} = \sum_{e_1,\dots,e_n} a_{e_1,\dots,e_n} \xi_{e_1}\cdots\xi_{e_n}$$

where $a_{e_1,\dots,e_n} = \int_{\mathrm{Tor}_n} e^{-2\pi i(e,\lambda)}\alpha(\lambda)\,d\lambda$. Then it is evident that the form

3.9. $$b^{(n)} = \sum_{e_1,\dots,e_n} b_{e_1,\dots,e_n} \xi_{e_1}\cdots\xi_{e_n} = \sum_{e_1,\dots,e_n} \xi_{e_1}\cdots\xi_{e_n} \sum_{m\in Z^1} a_{e_1+m,\dots,e_n+m}$$

belongs to $\mathcal{Y}_n^{(st)}$. From 3.9 we obtain the following representation for the coefficients of the form $b^{(n)}$

3.10. $$b_{e_1\dots e_n} = \int e^{-2\pi i(e,\lambda)} \,\mathcal{J}_p\Big(\sum_{r=1}^{n}\lambda_r\Big)\prod_{r=1}^{n} d\lambda_r$$

This means that, if the functions α',α'' coincide on the diagonal of $\mathrm{Tor}_n^{(st)}$:
$\{\lambda : \sum_{r=1}^{n}\lambda_r = 0 \pmod 1\}$ then they generate the same form $b^{(n)} \in \mathcal{Y}_n^{(st)}$.

Let $\mathcal{Y}^{(st)}$ be the algebraic sum of the linear spaces $\mathcal{Y}_n^{(st)}$. Every element $a \in \mathcal{Y}^{(st)}$ may be represented as a sum of similar stationary forms $a^{(n)}$ of degree n, $a = \sum_n a^{(n)}$ where in this sum only a finite number of terms is different from zero.

D.3.3. The space $\mathcal{Y}^{(st)}$ will be the tangent space of the gaussian stationary distribution.

Differential of the renormalization group.

Let us now give some formal arguments to introduce the differential of the renormgroup which we will indicate as $\partial_K A^*(\alpha)$. In section 2 we have already given formally the relation between $A^*_K(\alpha)\mathcal{G}$ and $\mathcal{G}$ where G is a probability measure on R^{Z^1}. For defining the differential of the adjoint operator of the renormalization group acting on the space of probability measures let us consider two probability distributions which are "near", that is $\mathcal{G}_\alpha$, $\mathcal{G}^\varepsilon_\alpha$ introduced in 3.5., 3.6., then we have that they transform in the following way:

3.8. $$A^*_K(\alpha)\mathcal{G} = \tilde{\mathcal{G}}_\alpha(B) = \int_{\xi:\, A_K(\alpha)\xi \in B} d\mathcal{G}_\alpha(\xi),$$

$$A^*_K(\alpha)\,\mathcal{G}^\varepsilon_\alpha = \tilde{\mathcal{G}}^\varepsilon_\alpha(B) = \int_{\xi:\, A_K(\alpha)\xi \in B} (1-\varepsilon a^{(n)})\, d\mathcal{G}_\alpha(\xi)$$

where B is some set belonging to $\mathcal{B}_{R^{Z^1}}$.

Then the tangent vector in $\tilde{\mathcal{G}}_\alpha$ can be formally defined as

3.9. $$-\frac{1}{\varepsilon}\left(\mathcal{G}^\varepsilon_\alpha - \mathcal{G}_\alpha\right) \xrightarrow[\varepsilon\to 0]{} \int_{\xi:\, A_K(\alpha)\xi \in B} a^{(n)}\, d\mathcal{G}_\alpha(\xi) = E(a^{(n)} \mid \tilde{\xi})$$

where $\tilde{\xi}_1 = (A_K(\alpha)\xi)_1$. Thus $\partial_K A^*(\alpha)$ will be defined as the transformation on $\mathcal{Y}^{(st)}$ given by

D.3.4. 3.10. $$\partial_K A^*(\alpha)\, a^{(n)} = E(a^{(n)} \mid \tilde{\xi})$$

Now we will give more details about this definition.

For this aim we shall use the n-stochastic integral defined by Ito and also some concepts from gaussian dynamical systems (18), (19).

Let us consider a gaussian dynamical system

3.11. $$(R^{Z^1}, \mathcal{B}_{R^{Z^1}}, G_{d}, T)$$

where R^{Z^1} is the space of realizations of the gaussian random field $\{\xi_i\}_{-\infty}^{+\infty}$ is the gaussian stationary aut.prob.distribution defined in 3.4; T is the shift operator:

3.12. $$(T\xi)_i = \xi_{i+1}$$

Then, from the theory of stationary stochastic processes we have that it is possible to represent the automodel gaussian random field with the help of its spectral random measure:

3.13. $$\xi_t = \int_{-\frac{1}{2}}^{+\frac{1}{2}} e^{2\pi i t\lambda} Z(d\lambda)$$

where the random measure $Z(d\lambda)$ forms a continuous complex normal random measure (18) $Z=\{Z(\Delta), \Delta\in\mathcal{H}\}$ where $\mathcal{H}$ is the algebra of borel sets belonging to $[-\frac{1}{2}, +\frac{1}{2}]$.

The continuous complex normal random measure $\{Z(\Delta), \Delta\in\mathcal{H}\}$ is a system of random variables such that for every collection of sets $\Delta_1, \dots, \Delta_n \in \mathcal{H}$ the joint probability distribution of the random variables $Z(\Delta_1), \dots, Z(\Delta_n)$ is gaussian with the following moments:

3.14. $$E(Z(\Delta_i)) = 0 \quad E(Z(\Delta_i)\overline{Z(\Delta_j)}) = \int_{\Delta_1\cap\Delta_2} \rho(\lambda)\,d\lambda .$$

Let T^* be the adjoint of the operator T, that is T acts on $Z(\Delta_i)$: from 3.13. it is easy to see that

3.15. $$T^* Z(d\lambda) = e^{i\lambda} Z(d\lambda)$$

Let $(\Omega, \mathcal{F}, P)$ be the basic probability space where all the complex random variables $Z(\Delta)$ are defined.

D.3.5. We will say that a complex random variable is a Baire function (18) of $\{Z(\Delta), \Delta\in\mathcal{H}\}$ if it belongs to the minimal class B that satisfies one of the two conditions:

(B1) When f is a complex valued Baire function of n complex variables in the usual sense $f(Z(\Delta_1),\ldots,Z(\Delta_n))$ belongs to B .

(B2) If f_n is a sequence in B which is convergent for every $w\in\Omega$ then the limit belongs to B.

D.3.6. $L^2(Z)$ will indicate the totality of the Baire functions of Z belonging to $L^2(\Omega,\mathcal{F},P)$.

Let us consider the n-dimensional Ito integral constructed with the help of the continuous complex normal random measure $\{Z(\Delta),\Delta\in\mathcal{H}\}$

3.16. $$I_n(f)=\int_{-\frac{1}{2}}^{+\frac{1}{2}}\cdots\int_{-\frac{1}{2}}^{+\frac{1}{2}} f(\lambda_1,\ldots,\lambda_n)\, Z(d\lambda_1)\cdots Z(d\lambda_n)$$

where $f(\lambda_1,\ldots,\lambda_n)$ is such that

3.17. $$\int |f(\lambda_1,\ldots,\lambda_n)|^2 g(\lambda_1)\cdots g(\lambda_n)\, d\lambda_1\cdots d\lambda_n<+\infty$$

and the region of integration in 3.16 does not contains points $(\lambda_1,\ldots,\lambda_n)$ such that $\sum_i\lambda_i=0$ (19).

From the theory of Ito integrals (18) we have the following propositions useful for our aims:

Let us define the scalar product between two random variables ξ,η using the basic probability space $L^2(\Omega,\mathcal{F},P)$

3.18. $$(\xi,\eta)=\int_\Omega \xi(\omega)\eta(\omega)\,dP$$

Then we have

Proposition 3.1. $$(I_n(f),I_m(g))=\delta_{n,m}\int_{-\frac{1}{2}}^{+\frac{1}{2}} f(\lambda_1,\ldots,\lambda_n)\bar{g}(\lambda_1,\ldots,\lambda_n)\prod_i g(\lambda_i)\,d\lambda_i$$

Proposition 3.2. The system

$$\{I_n(f),\ n=0,1,\cdots\}$$

where f satisfies 3.17. is complete in $L^2(Z)$

Proposition 3.3. $$\int_{-1/2}^{+\frac{1}{2}}\cdots\int_{-1/2}^{+1/2} f(t_1)\cdots f(t_n)\,Z(dt_1)\cdots Z(dt_n)=H_n(\mathcal{Z})$$

where $\mathcal{Z}=\int f(t)Z(dt)$ and $H_n(x)$ is an hermite polynomial of the real variable x.

Now we will choose as base in the space $\mathcal{Y}^{(st)} \subset L^2(Z)$ the hermite polynomials and we will find an explicit expression for 3.10.

Lemma 3.1. Let $\tilde{\mathcal{B}}_{R^{z^1}}$ be the σ- algebra generated by the random variables $\{\tilde{\xi}_r\}_{-\infty}^{+\infty}$ defined as in 3.9. for some fixed k. Then the conditional mathematical expectation $E(I_m(f)|\tilde{\mathcal{B}}_{R^{z^1}})$ will be given by

3.19. $$E(I_m(f)|\tilde{\mathcal{B}}_{R^{z^1}}) = I_m(g) = \int\cdots\int g(\lambda_1,\dots,\lambda_n)\tilde{Z}(d\lambda_1)\cdots\tilde{Z}(d\lambda_n)$$

where $\tilde{Z}(d\lambda)$ is the spectral random measure of $\{\tilde{\xi}_i\}_{-\infty}^{+\infty}$ and

$$g(\lambda_1,\dots,\lambda_m) = K^{-m(\frac{d}{2}+1)} \sum_{K\mu_j=\lambda_j\,(\mathrm{mod}\,1),\, j=1,\dots,n} f(\mu_1,\dots,\mu_m) \prod \rho_1(\mu_j)\rho_1^{-1}(K\mu_j)$$

and $\rho_1(\lambda) = \rho(\lambda)/(e^{2\pi i\lambda}-1)$ being given by (2.18.)

Proof. We will show the case n=1. We have

3.20. $$\tilde{\xi}_r = \frac{1}{K^{d/2}} \int_{-\frac{1}{2}}^{+\frac{1}{2}} \sum_{s=rK}^{(r+1)K-1} e^{2\pi i s\lambda} Z(d\lambda) =$$

$$= K^{-\frac{d}{2}} \int_{-\frac{1}{2}}^{+\frac{1}{2}} e^{2\pi i\lambda rK}(1+e^{2\pi i\lambda}+\dots+e^{2\pi i\lambda(K-1)})\dot{Z}(d\lambda) =$$

$$= K^{-\frac{d}{2}} \int_{-\frac{1}{2}}^{+\frac{1}{2}} e^{2\pi i\lambda rK}(e^{2\pi i\lambda K}-1)(e^{2\pi i\lambda}-1)^{-1} Z(d\lambda)$$

setting $\lambda K = \lambda'$ we can compare this expression with $\tilde{\xi}_r = \int_{-\frac{1}{2}}^{+\frac{1}{2}} e^{2\pi i\lambda r}\tilde{Z}(d\lambda)$ and thus

3.21. $$\tilde{\xi}_r = K^{-\frac{d}{2}} \int_{-\frac{1}{2}}^{+\frac{1}{2}} e^{2\pi i\lambda r}(e^{2\pi i\lambda}-1) \sum_{c=0}^{K-1} \frac{1}{e^{2\pi i(\frac{\lambda'}{K}+\frac{c}{K})}-1} Z(d\lambda')$$

and we obtain the following relation between $\tilde{Z}(d\lambda)$ and $Z(d\lambda)$

3.22. $$\int_\Delta \frac{\tilde{Z}(d\lambda)}{e^{2\pi i\lambda}-1} = K^{-\frac{d}{2}} \sum_{\Delta' : K\Delta' \equiv \Delta\,(\mathrm{mod}\,1)} \int_{\Delta'} \frac{Z(d\lambda)}{e^{2\pi i\lambda}-1}$$

We now resolve (3.22.). Let us call S_K the transformation on intervals used in (3.22.)

3.23. $$S_K \lambda = K\lambda \pmod 1$$

We can find a function $C(\mu)$ such that

3.24. $$\int_{\Delta} Z(d\lambda)\,(e^{2\pi i\lambda}-1)^{-1} = \int_{S_K\Delta} C(\mu)\,\tilde{Z}(d\mu)\,(e^{2\pi i\mu}-1)^{-1} + Z'(\Delta)$$

where $Z'(\Delta)$ must be independent from $\tilde{Z}(\Delta)$ for every $\Delta \subset [-\frac{1}{2}, +\frac{1}{2}]$. Let us make the scalar product of 3.24. with $\int_{\tilde{\Delta}} \tilde{Z}(d\lambda)(e^{2\pi i\lambda}-1)^{-1}, \tilde{\Delta} \subset S_K\Delta$ from the above condition we have:

3.25. $$\left(\int_{\tilde{\Delta}} \frac{\tilde{Z}(d\lambda)}{e^{2\pi i\lambda}-1}, \int_{\Delta} \frac{Z(d\lambda)}{e^{2\pi i\lambda}-1}\right) = \left(\int_{\tilde{\Delta}} \frac{\tilde{Z}(d\lambda)}{e^{2\pi i\lambda}-1}, \int_{S_K\Delta} \frac{C(\mu)\tilde{Z}(d\mu)}{e^{2\pi i\mu}-1}\right)$$

the left hand side of 3.25. can be computed using 3.22. and obtain

3.26. $$l.h.s = K^{-\frac{\alpha}{2}} \int_{\Delta\cap S_K^{-1}\tilde{\Delta}} \frac{\rho(\lambda)d\lambda}{|e^{2\pi i\lambda}-1|^2}$$

where we have choosen Δ in such a way that Δ intersects only one of the $S_K^{-1}(S_K\Delta)$

Thus, setting $\tilde{\Delta}' = \Delta \cap S_K^{-1}\tilde{\Delta}$, and evaluating the scalar product on the right hand side:

3.27. $$K^{-\frac{\alpha}{2}-1} \int_{\tilde{\Delta}'} \rho(\lambda)\,|e^{2\pi i\lambda}-1|^{-2} d\lambda = \int_{\tilde{\Delta}} C(\mu)\rho(\mu)\,|e^{2\pi i\mu}-1|^{-2} d\mu$$

from which we obtain

3.28. $$C(\mu) = K^{-\frac{\alpha}{2}-1} \frac{\rho(\lambda)}{\rho(\mu)} \frac{|e^{2\pi i\mu}-1|^2}{|e^{2\pi i\lambda}-1|^2}, K\lambda \equiv \mu \pmod 1$$

Introducing the new random measures (20)

3.29. $$\psi(\Delta) = \int_{\Delta} \rho_1^{-1}(\lambda) Z(d\lambda), \quad \tilde{\psi}(\Delta) = \int_{\Delta} \rho_1^{-1}(\lambda) Z(d\lambda)$$

it is possible to write 3.24. in a very simple form;

3.30. $$\Psi(\Delta)=K^{-\frac{\alpha}{2}-1}\tilde{\Psi}(S_K\Delta)+\Psi'(\Delta)$$

where $\Psi'(\Delta)$ is independent with respect to all the $\tilde{\Psi}(\Delta), \Delta\subset[-\frac{1}{2},+\frac{1}{2}]$ and $E\Psi'(\Delta)=0$

Let us now proof the lemma for $n=1$; Let us divide the interval $[-\frac{1}{2},+\frac{1}{2}]$ in disjoint intervals of equal length $\Delta_e=\left(\frac{e}{K^m},\frac{e+1}{K^m}\right)$, $m>0$ and let us construct the Ito integral for $n=1$ with a function g defined by $g_t=a_e$ if $t\in\Delta_e$. Then

3.31. $$I_1(g)=\Sigma_e a_e Z(\Delta_e)=\Sigma_e a_e \rho_1(\bar{\lambda}_e)\Psi(\Delta_e)$$

where $\bar{\lambda}_e\in\Delta_e$. Then

3.32. $$E(I_1(g)|\tilde{\mathcal{B}}_{R^{Z^1}})=\Sigma_e\frac{a_e\rho_1(\bar{\lambda}_e)}{K^{\alpha/2+1}}E(\tilde{\Psi}(S_K\Delta_e)|\tilde{\mathcal{B}}_{R^{Z^1}})=$$

$$=K^{-\frac{\alpha}{2}-1}\Sigma_e a_e\rho_1(\bar{\lambda}_e)\tilde{\Psi}(S_K\Delta_e)=$$

$$=K^{-\frac{\alpha}{2}-1}\sum_{e'}\tilde{\Psi}(\Delta_{e'})\sum_{e:S_K\Delta_e=\Delta_{e'}}a_e\rho_1(\bar{\lambda}_e)=$$

$$=K^{-\frac{\alpha}{2}-1}\Sigma_{e'}\frac{\tilde{Z}(\Delta_{e'})}{\rho_1(K\bar{\lambda}_e)}\sum_{e:S_K\Delta_e=\Delta_{e'}}a_e\rho_1(\bar{\lambda}_e)=$$

$$=I_1(g)$$

where $$\Delta_{e'}=\left(\frac{e}{K^{m-1}},\frac{e+1}{K^{m-1}}\right)$$

Now we can give the fundamental result found by Ja.G.Sinai (12).

Set

3.33. $$\varphi_n(\lambda_1,\dots,\lambda_n)=\prod_{j=1}^{n}\rho_1^{-1}(\lambda_j)\sum_{M}\prod_{j=1}^{n}\frac{sgn(m_j+\lambda_j)}{|m_j+\lambda_j|^{\alpha}}$$

where $\sum_{M}$ means sum over all the n-uples such that $\sum_{j=1}^{n}(\lambda_j+m_j)=0$.

Then we can establish

Theorem 3.1. Let $\chi(\lambda)$ be a symmetric C^1 function defined on the dimensional torus, $\chi(\lambda)=\varphi_n(\lambda)$, for $\lambda \in \mathrm{Tor}_n^{st}$.

Let us consider the Ito integral $I_n(\Psi)$. Then

1) the form $h^{(n)}=\sum_{m\in Z^1}(T_m^*)I_m(\Psi)\in \mathcal{Y}^{(st)}$ is an eigenfunction for $\{\partial_K A^*(\alpha)\}$ and we have that

$$\partial_K A^*(\alpha)\, h^{(n)} = K^{\gamma_n} h^{(n)},\ \gamma_n=\frac{n\alpha}{2}-n+1$$

2) If $a^{(n)}=\sum_{m\in Z^1}(T_m^*)(I_n(f))\in \mathcal{Y}^{(st)}$

where f is a symmetrical C^1 function defined on $\mathrm{Tor}n^{(st)}$ then

$$\lim_{K\to\infty} K^{-\gamma_n}\partial_K A^* a^{(n)} = (2\pi i)^n f(0,\ldots,0)\, h^{(n)}$$

From theorem 3.1. it is evident that for n=4 we expect a bifurcation for $\alpha=\frac{3}{2}$. In the following lectures we shall give some methods for finding non gaussian automodel probability distributions.

4. Eigenfunction and eigenvalue of the differential defined for a Gaussian automodel probability distribution and ε-expansion *)

4.1. As we have seen in the preceeding lectures we have a new branch of non gaussian automodel distributions for values of α equal to $\frac{3}{2}$ corresponding to the case n=4. In this lecture we will give an explicit construction of eigen-

*) The content of this lecture is exposed in a not yet published paper of P.M. Bleher (21).

functions and eigenvalues of the differential of the renormalization group in such a way that it will be possible to determine the hamiltonian Heff of this new non gaussian automodel distributions using an expansion method. The system of recurrent equations thus obtained are resolved up to the second corder.

From the results obtained it is evident that the nongaussian terms correspond to a decrease of the potential which is slower than that of the gaussian term: this question can be further investigated resolving the third order or by applying the methods here developed.

We will make use of the random process (generalized) introduced in the first lecture

4.1. $$\theta(t) = \frac{1}{e^{-it}-1}\,\eta(t)$$

where $\eta(t)$ is a periodic generalized stationary gaussian process defined as the Fourier transform of the automodel Gaussian field $\{\xi_i\}_{-\infty}^{+\infty}$. We have already determined the correlation function of $\theta(t)$ in the preeceding lectures where we have found that it is equal to

4.2. $$\langle \theta(t_1)\theta(t_2)\rangle = g(t_1)\delta_p(t_1+t_2) = \frac{\rho(t_1)}{|e^{2\pi i t_1}-1|^2}\,\delta_p(t_1+t_2)$$

where $\rho(t)$ is the spectral function of the gaussian automodel stationary field $\{\xi_n\}_{-\infty}^{+\infty}$.

We will write a stationary n-form as

4.3. $$H(\theta) = \int_{-\pi}^{+\pi}\cdots\int_{-\pi}^{+\pi} f(t_1,\dots,t_n)\,\theta(t_1)\cdots\theta(t_n)\,dt_1\cdots dt_n$$

where $f(t_1,\dots,t_n) = \delta_p(t_1+\dots+t_n)\,g(t_1,\dots,t_n)$

It is possible, but difficolt, to show the equivalence between this kind of approach and the one obtained by using the Ito integral and we assume that this fact is true *)

So we assume that it is possible to make the formal change between $\theta(t)$ and

$Z(d\lambda)$ and then we find the relation between $\theta(t)$ and $\theta'(t)$ analogous to the one between $Z(d\lambda)$, $Z'(d\lambda)$

4.4. $$\theta(t) = a(t)\,\theta'(Kt) + \zeta(t)$$

where $\zeta(t)$ is a gaussian random process independent with respect to all the $\theta'(t)$ for every t and $a(t)$ is given by:

4.5. $$a(t) = \frac{G(t)}{K^{\alpha/2}\, G'(Kt)}$$

where G' is the correlation function of the process $\theta'(t)$.

4.2. Explicit expression for $\partial_K A^*(\alpha)$.

For finding how the n-form (4.3.) transforms under the action of the differential $\partial_K A^*(\alpha)$ we have to calculate the conditional expectation of the variables $H(\theta(t))$ under the conditions θ' fixed. Thus

4.6. $$H'(\theta') = E(H(\theta)|\theta') = \int_{-\pi}^{+\pi}\cdots\int_{-\pi}^{+\pi} f(t_1,\dots,t_n)\cdot$$
$$\cdot E\Big(\prod_{j=1}^{n} a(t_j)\,\theta'(Kt_j) + \zeta(t_j)\,|\,\theta'\Big)dt_1\cdots dt_n$$

we have that $E'(\theta'|\theta') = \theta'$ and since $\zeta(t_j)$ is independent from the θ' we have that $E(\zeta(t_i)|\theta') = E\,\zeta(t_j)$ and the contribute to (4.6.) different from zero will come only from the terms which contain an even number of $\zeta(t_j)$ and so we have only to calculate the mean of products of $\zeta(t_j)$ and this can be made using the Wick expansion and the correlation function $\langle\zeta(t_i)\,\zeta(t_j)\rangle$ which can be obtained by (4.3.).

---

*) F.Dinaburg and Ja.G.Sinai have shown that is possible to represent the n-form using the same functions as here but through stochastic integrals. (Unpublished result).

So we have

4.7. $$H'(\theta') = \overline{H}_n(\theta) + \overline{H}_{n-2}(\theta) + \cdots$$

By setting $\partial_1 A^*(\alpha) : H_n(\theta) \to \overline{H}_n(\theta)$ we can find immediately how it acts, in fact we have:

4.8. $$[H'(\theta')]_n = \int_{-\pi}^{+\pi} \cdots \int_{-\pi}^{+\pi} f(t_1,\ldots,t_n) \prod_{j=1}^{n} a(t_j)\, \theta'(Kt_j)\, dt_1 \cdots dt_n$$

making the change of variables $Kt_j \to t_j$ we obtain:

4.9. $$[H'(\theta')]_n = \int_{-\pi}^{+\pi} \cdots \int_{-\pi}^{+\pi} K^{-n} \sum_{j_1,\ldots,j_n=1}^{K} f\left(\frac{t_1}{K} + \frac{2\pi j_1}{K}, \ldots, \frac{t_n}{K} + \frac{2\pi j_n}{K}\right) \cdot$$

$$\cdot\, a\left(\frac{t_1}{K} + \frac{2\pi}{K} j_1\right) \cdots a\left(\frac{t_n}{K} + \frac{2\pi}{K} j_n\right) \cdot \theta'(t_1) \cdots \theta'(t_n)\, dt_1 \cdots dt_n$$

Using the expression (4.5.) for a (t) and writing

4.10. $$[H'(\theta')]_n = \int_{-\pi}^{+\pi} \cdots \int_{-\pi}^{+\pi} f(t'_1,\ldots,t'_n)\, \theta'(t_1) \cdots \theta'(t_n)\, dt_1 \cdots dt_n$$

we have

4.11. $$f'(t_1,\ldots,t_n) = K^{-\frac{n\alpha}{2}}\, T_K\left[f_n(t_1,\ldots,t_n) \frac{G(t_1)\cdots G(t_n)}{G(Kt_1)\cdots G(Kt_n)} \right]$$

where T_K is the mean on the torus Torn:

4.12. $$T_K[g(t_1,\ldots,t_n)] = K^{-n} \sum_{j_1,\ldots,j_n=1}^{K} g\left(\frac{t_1}{K} + \frac{2\pi}{K} j_1, \ldots, \frac{t_n}{K} + \frac{2\pi}{K} j_n\right)$$

Setting $g(t_1,\ldots,t_n) = f(t_1,\ldots,t_n) G(t_1) \cdots G(t_n)$ we can find the eigenvectors of $\partial_1 A^*(\alpha)$ if we find the eigenfunction of the equation

4.13. $$g'_n(t_1,\ldots,t_n) = K^{-\frac{n\alpha}{2}}\, T_K[g(t_1,\ldots,t_n)]$$

4.13.

4.3. Eigenfunctions of $\partial_1 A^*(\alpha)$

Let us look for the eigenfunctions of T_K . We choose the initial functions within the set of functions which have the following form

4.14. $$Q_0(t_1,\dots,t_n)=\delta_p(t_1+\dots+t_n)Q(t_1,\dots,t_n)=(2\pi)^{-1}\sum_{\tau=-\infty}^{+\infty} e^{i\tau(t_1+\dots+t_n)}Q(t_1,\dots,t_n)$$

for $|t_i|\le\pi$, where Q is a homogeneus function of degree $(-m)$. The eigenfunction of T_K can be found using the limit $T_K Q_0 \xrightarrow[K\to\infty]{} \bar{Q}$.

We have:

$$T_K[Q_0(t_1,\dots,t_n)]=(2\pi)^{-1}\sum_{\tau=-\infty}^{+\infty} T_K[e^{i\tau(t_1+\dots+t_n)}Q(t_1,\dots,t_n)]=$$

4.15. $$=(2\pi)^{-1}K^{-n}\sum_{j_1,\dots,j_n=1}^{K}\sum_{\tau=-\infty}^{+\infty} e^{i\frac{\tau}{K}(t_1+\dots+t_n)}K^{m}Q(t_1+2\pi j_1,\dots,t_n+2\pi j_n)$$

For $K\to\infty$ we can set $\sum_{\tau=-\infty}^{+\infty} K^{-1}F\left(\frac{\tau}{K}\right)\to\int_{-\infty}^{+\infty}F(\tau)\,d\tau$

If $m>n-1$ we have that $Q(t_1+2\pi j_1,\dots,t_n+2\pi j_n)$ is not integrable on the surface $t_1+2\pi j_1+\dots+t_n+2\pi j_n=0$ and so we obtain $\lim_{K\to\infty} K^{-m+n-1}T_K[Q_0(t_1,\dots,t_n)]=$

4.16. $$=(2\pi)^{-1}\int_{-\infty}^{+\infty}d\tau\sum_{j_1\dots j_n=-\infty}^{+\infty} e^{i\tau(t_1+2\pi j_1+\dots+t_n+2\pi j_n)}Q(t_1+2\pi j_1,\dots,t_n+2\pi j_n)=$$

$$=\sum_{j_1,\dots,j_n=-\infty}^{+\infty}\delta_p(t_1+2\pi j_1,\dots,t_n+2\pi j_n)\,Q(t_1+2\pi j_1,\dots,t_n+2\pi j_n)$$

If we set

4.17. $$U_n[g(t_1,\dots,t_n)]=\sum_{j_1,\dots,j_n=-\infty}^{+\infty} g(t_1+2\pi j_1,\dots,t_n+2\pi j_n)$$

then we can say that $\int_{-\infty}^{+\infty}d\tau\,U_n[e^{i\tau(t_1+\dots+t_n)}Q(t_1,\dots,t_n)]$

is an eigenfunction with eigenvalue $\lambda_n^{(m)}=K^{-\frac{nd}{2}+m-n+1}$

For $m<n-1$ we have that $T_K[Q_0(t_1,\dots,t_n)]$ tends to integral over the torus Torn thus the function

4.18. $$g_0(t_1,\dots,t_m) = \delta_p(t_1+\dots+t_n)$$

is eigenfunction of T_k.

Now we are able to find the eigenfunctions of eq. (4.11.) using the fact that the $G(t)$ are invariant with respect the action of the renormalization group:

Proposition 4.1. The eigenfunctions of the transformation (4.11.) are given by

1) for $m>n-1$

$$f_m(t_1,\dots,t_m) = \prod_{i=1}^{n} G(t_i)^{-1} \sum_{j_1,\dots,j_n : \sum_1^n e(t_e+2\pi j_e)=0} Q(t_1+2\pi j_1,\dots,t_n+2\pi j_n)$$

with eigenvalue $K^{-\frac{n\alpha}{2}+m-n+1}$

2) there is also another eigenfunction $f_m(t_1,\dots,t_m) = \delta_p(t_1+\dots+t_n) G^{-1}(t_1)\dots G^{-1}(t_m)$

Remark. Note that the global transformation $\partial_k A^*(\alpha)$ is triangular so that its global properties are determined by the eigenvalues and the eigenfunctions of the transformation $\partial_1 A^*(\alpha)$. We can also note that it is possible to find a larger class of eigenfunctions if we allow the $f_m(t_1,\dots,t_m)$ to be generalized functions.

Now we make a selection within the set of the eigenfunction of proposition (4.1.), because we want that the effective hamiltonian of degree n

4.19. $$H_m(\eta) = \int_{-\pi}^{+\pi}\dots\int_{-\pi}^{+\pi} e_m(t_1,\dots,t_m)\,\eta(t_1)\dots\eta(t_m)\,dt_1\dots dt_m$$

to satisfy some physical properties, where it is easy that e_m is connected with f_m of prop. (4.1.) by the following equation:

4.20. $$e_m(t_1,\dots,t_m) = \delta_p(t_1+\dots+t_m) \prod_{i=1}^{n} G(t_i)^{-1}(e^{-2\pi i t_i}-1)^{-1} \cdot \sum_{j_1,\dots,j_m} Q(t_1+2\pi j_1,\dots,t_m+2\pi j_m)$$

The required properties are

1) The effective hamiltonian must be translation invariant

2) H_n must be positive definite in order to generate a Gibbs measure

3) the potential corresponding to the hamiltonians with more than two particles must decrease faster than the potential coming from the gaussian term.

Thus

4.12. $$\prod_{i=1}^{n} G(t_i)^{-1}(e^{-i2\pi t_i}-1)^{-1}\sum_{j_1,\dots,j_n=-\infty}^{+\infty} Q(t_1,\dots,t_n)\neq 0, <\infty$$

$$\text{for } t_1,\dots,t_n=0$$

From (4.21.) it is evident that the singularity of Q in the origin must cancel with the behaviour of $\prod_i G(t_i)^{-1}(e^{-2\pi i t_i}-1)^{-1}$

Let us evaluate $G(t)$ in the origin

4.22. $$G(t_j)=\sum_{e=-\infty}^{+\infty}\frac{1}{|t_j+2\pi e|^{1+\alpha}},\quad G(t)\sim |t|^{-1-\alpha}$$

Thus $G(t)(e^{-it}-1)\sim -it|t|^{-1-\alpha}$ and

4.23. $$Q(t_1,\dots,t_n)=\prod_{j=1}^{n}\tilde{\rho}(t_j)\qquad \tilde{\rho}(t)=-it|t|^{-1-\alpha}$$

thus we find the expression for the eigenfunction of $\partial_1 A^*(\alpha)$

4.24. $$e_n(t_1,\dots,t_n)=\Big[\prod_{j=1}^{n}(e^{-it_j}-1)G(t_j)\Big]^{-1}\cdot\Big(\sum_{j_1,\dots,j_n,\ \sum_{k=1}^{n}(t_k+2\pi j_k)=0}\ \prod_{k=1}^{n}\tilde{\rho}(t_k+2\pi j_k)\Big)$$

which have already been determined in (12).

4.4. Compact formulae for the full differential.

Now we are able to give a representation of the action of the full differential $\partial_K A^*(\alpha)$.

We shall see how it acts on the space of forms of the type

4.25. $$H_n(\eta) = \int_{-\pi}^{+\pi}\cdots\int_{-\pi}^{+\pi} e_n(t_1,\ldots,t_n)\,\eta(t_1)\cdots\eta(t_n)\,dt_1\cdots dt_n$$

where $e_n(t_1,\ldots,t_n)$ is given by (4.24.). It is useful to use now other random variables defined by

4.26. $$w(t) = \frac{1}{e^{-it}-1}\,\frac{1}{g(t)}\,\eta(t)$$

The transformation induced on $W(t)$ by the action of $A_K(\alpha)$ is determined by the preceeding formulas (4.4.), (4.5.)

4.27. $$w(t) = K^{-\frac{\alpha}{2}} w'(Kt) + \chi(t)$$

where $\chi(t)$ is a complex gaussian random process whose correlation function is determined by that of $w(t), w'(Kt)$ and is independent from $w'(Kt)$. Using $w(t)$ we have another form for $H(\eta)$

4.28. $$H(w) = \int_{-\pi}^{+\pi}\cdots\int_{-\pi}^{+\pi} g_n(t_1,\ldots,t_n)\,w(t_1)\cdots w(t_n)\,dt_1\cdots dt_n$$

where $g_n(t_1,\ldots,t_n)$ is given by

4.29. $$g_n(t_1,\ldots,t_n) = \sum_{\substack{j_1,\ldots,j_n \\ \Sigma_K (t_K+2\pi j_K)=0}} \prod_{K=1}^{n} \tilde{\varsigma}(t_K+2\pi j_K)$$

Thus

4.30. $$H(w) = (2\pi)^{-1}\int_{-\infty}^{+\infty} dz \int_{-\pi}^{+\pi} dt_1 \cdots \int_{-\pi}^{+\pi} dt_n\, w(t_1)\cdots w(t_n)\,\cdot$$
$$\cdot\, U_n\Big[\prod_{K=1}^{n} \tilde{\varsigma}(t_K)\,e^{it_K z}\Big] = (2\pi)^{-1}\int_{-\infty}^{+\infty} dz \left(\int_{-\pi}^{+\pi} dt\, w(t)\, U_1[\tilde{\varsigma}(t)\,e^{itz}]\right)^{n}$$

Let us define the function

4.31. $$P_\alpha(\eta,t)=(-i)\sum_j \frac{(t+2\pi j)}{|t+2\pi j|^{1+\alpha}} e^{i\eta(t+2\pi j)}$$

So that the eigenvector of $\partial_1 A^*(\alpha)$ will have the form

4.32. $$H_n(w)=(2\pi)^{-1}\int_{-\infty}^{+\infty} d\eta\left(\int_{-\pi}^{+\pi} dt\, w(t)\, P_\alpha(\eta,t)\right)^n$$

We can determine now how $\partial_K A^*(\alpha)$ acts on the more general form:

4.33. $$H(w(t))=\int_{-\infty}^{+\infty}(2\pi)^{-1} d\eta\, F\left(\int_{-\pi}^{+\pi} dt\, w(t)\, P_\alpha(\eta,t)\right)$$

where $F(x)$ is some polynomial. We have

4.34. $$\partial_K A^*(\alpha)\, H(w(t))=E(H|w'(t))$$

We calculate (4.34.) in this way. Set $x=\int_{-\pi}^{+\pi} dt\, P_\alpha(\eta,t)\, w(t)$, from well known theorems (18) x will be a complex gaussian random process whose variation is determined by that of $w(t)$, and $w'(t)$:

4.35. $$Ex=0,\; E|x|^2=E\left|\int_{-\pi}^{+\pi} P_\alpha(\eta,t)\, w(t)\, dt\right|^2=\gamma_0(\eta)$$

Thus we can set:

$$E(H(w)|w'(t))=E(E(H(w)|w'(t))|x)=$$

$$=E(E(\int_{-\infty}^{+\infty}(2\pi)^{-1}d\eta\, F(\int_{-\pi}^{+\pi}dt\, K^{-\frac{\alpha}{2}}w'(Kt)P_\alpha(\eta,t)+\int_{-\pi}^{+\pi}dt\, x(t)\, P_\alpha(\eta,t)|w'(t)|$$

$$|x)=$$

4.36. $$=\int_{-\infty}^{+\infty} d\eta\,(2\pi\gamma_0(\eta))^{-\frac{1}{2}}\int_{-\infty}^{+\infty} dx\, e^{-\frac{x^2}{2\gamma_0(\eta)}} F\left(K^{-\frac{\alpha}{2}}\int_{-\pi}^{+\pi} P_\alpha(\eta,t)\, w'(Kt)\, dt+x\right)=$$

~~$$=K^{-1-\frac{\alpha}{2}+\alpha}\int_{-\pi}^{+\pi} P_\alpha\left(\frac{\eta}{J},t\right) w'(t)\, dt$$~~

Let us make the substitution $Kt \to t$:

4.37. $$K^{-1-\frac{\alpha}{2}} \int_{-\pi K}^{+\pi K} P_\alpha(\eta, \tfrac{t}{K}) w'(t) dt = K^{-1-\frac{\alpha}{2}} \int_{-\pi}^{+\pi} \sum_{j=0}^{K-1} P_\alpha(\eta, \tfrac{t}{K} + \tfrac{2\pi}{K} j) w'(t) dt =$$

$$= K^{-1-\frac{\alpha}{2}+\alpha} \int_{-\pi}^{+\pi} P_\alpha(\tfrac{\eta}{j}, t) w'(t) dt$$

Since $\sum_{j=1}^{K} P_\alpha(\eta, \frac{t}{K} + \frac{2\pi}{K} j) = K^\alpha P_\alpha(\frac{\eta}{K}, t)$, we have:

4.38. $$H'(w') = \int_{-\infty}^{+\infty} d\eta \, (2\pi \gamma_0(\eta))^{-\frac{1}{2}} \int_{-\infty}^{+\infty} dx \, e^{-\frac{x^2}{2\gamma_0(\eta)}} F\left(K^{\frac{\alpha}{2}-1} \int_{-\pi}^{+\pi} P_\alpha(\tfrac{\eta}{K}, t) w'(t) dt + x\right)$$

changing $\eta/K \to \eta$ and setting $\gamma(\eta) = \gamma_0(K\eta)$ we obtain a trasformation formula for F :

4.39. $$F'(z, \eta) = \frac{K}{\sqrt{2\pi\gamma(\eta)}} \int_{-\infty}^{+\infty} dx \, e^{-\frac{x^2}{2\gamma(\eta)}} F\left(\frac{z}{K^{1-\frac{\alpha}{2}}} + x\right)$$

Choosing $F(x)$ to be a polynomial in the variable z , with coefficients depending on η we can rewrite (4.39.) in the final form:

4.40. $$F'(z', \eta') = \frac{K}{\sqrt{2\pi\gamma(\eta)}} \int_{-\infty}^{+\infty} dx \, e^{-\frac{x^2}{2\gamma(\eta)}} F\left(\frac{z}{\sqrt{c}} + x, K\eta\right)$$

where $C = K^{2-\alpha}$. In this way we have reduced the problem of the action of the differential of the renormalization group $\partial_K A^*(\alpha)$ to the problem of studying the properties of the integral operator (4.40.).

5. Eigenfunction of the differential of the renormalization group.

It is possible to study the properties of the operator (4.40.). First we shall evaluate $\gamma(\eta)$:

$$\gamma(\ell)=\gamma_0(K\ell)=\left\langle\left|\int_{-\pi}^{+\pi}P_\alpha(K\ell,t)\chi(t)dt\right|^2\right\rangle=$$

4.41.
$$=\int_{-\pi}^{+\pi}dt_1\int_{-\pi}^{+\pi}dt_2\,P_\alpha(K\ell,t_1)\overline{P_\alpha(K\ell,t_2)}\langle\chi(t_1)\overline{\chi(t_2)}\rangle$$

From the definition (4.27.) of $\chi(t)$ we have that

$$\langle\chi(t_1)\overline{\chi(t_2)}\rangle=\langle\chi(t_1)\overline{\omega(t_2)}\rangle=\langle(\omega(t_1)-K^{-\frac{\alpha}{2}}\omega'(Kt_1))\overline{\omega(t_2)}\rangle=$$

4.42.
$$=\frac{1}{G(t_1)G(t_2)}G(t_2)\delta_p(t_1+t_2)-\frac{K^{-\frac{\alpha}{2}-1-\frac{\alpha}{2}}}{G(Kt_1)G(Kt_2)}\sum_{j=1}^{K}\langle\theta(t_1+\tfrac{2\pi}{K}j)\overline{\theta(t_2)}\rangle=$$

$$=G(t_1)^{-1}\delta_p(t_1+t_2)-K^{-1-\alpha}G^{-1}(Kt_1)\sum_{j=1}^{K}\delta_p(t_1+t_2+\tfrac{2\pi}{K}j)$$

Putting (4.42.) into (4.41.) we obtain:

$$\gamma(\ell)=\int_{-\pi}^{+\pi}dt_1\int_{-\pi}^{+\pi}dt_2\,P_\alpha(K\ell,t_1)\overline{P_\alpha(K\ell,t_2)}G(t_1)^{-1}\delta_p(t_1+t_2)+$$

4.43.
$$-\int_{-\pi}^{+\pi}dt_1\int_{-\pi}^{+\pi}dt_2\,P_\alpha(K\ell,t_1)\overline{P_\alpha(K\ell,t_2)}\cdot G(Kt_1)^{-1}\cdot$$

$$\cdot K^{-1-\alpha}\sum_{j=1}^{K}\delta_p(t_1+t_2+\tfrac{2\pi}{K}j)$$

4.44.
$$S_1=-\int_{-\pi}^{+\pi}dt_1\int_{-\pi}^{+\pi}dt_2\sum_{j_1,j_2=-\infty}^{+\infty}\frac{(t_1+2\pi j_1)}{|t_1+2\pi j_1|^{1+\alpha}}\frac{(t_2+2\pi j_2)}{|t_2+2\pi j_2|^{1+\alpha}}\cdot$$

$$\cdot\frac{\delta_p(t_1+t_2)}{G(t_1)}e^{iK\ell(t_1+t_2+2\pi j_1+2\pi j_2)}$$

Setting $\quad \rho_0(t)=\dfrac{t}{|t|^{1+\alpha}}$

$$S_1=\int_{-\infty}^{+\infty}\int_{-\infty}^{+\infty} dt_1 dt_2\, \rho_0(t_1)\rho_0(t_2)\, e^{ik\tau(t_1+t_2)} g^{-1}(t_1)\delta_p(t_1+t_2)=$$

4.45. $$=\sum_{\ell=-\infty}^{+\infty}\int_{-\infty}^{+\infty} dt\, \rho_0(t)\, \rho_0(2\pi\ell-t)\, e^{ik\tau 2\pi\ell} g^{-1}(t)dt=\psi(K\tau)$$

where $\psi(\tau)=-\sum_{\ell=-\infty}^{+\infty}\int_{-\infty} dt\, \frac{\rho_0(t)\rho_0(2\pi\ell-t)}{g(t)} e^{i2\pi\ell\tau}$ in a similar way it is possible to show that $S_2=K^{\alpha-2}\psi(\tau)$ and so we obtain

4.46. $$\gamma(\tau)=\psi(K\tau)-K^{\alpha-2}\psi(\tau)$$

Now it is possible to find the explicit expressions of the eigenfunctions of $\partial_K A^*(\alpha)$. Let us define before the hermite polynomials:

$$h_n(z;a)=z^n-\frac{1}{2}\binom{n}{2}az^{n-2}+\frac{13}{2^2}\binom{n}{4}a^2z^{n-4}+\dots$$

4.47. $$\int_{-\infty}^{+\infty} h_n(z;a)\, h_m(z;a)\exp\left(-\frac{z^2}{a}\right)dz=0\ , \text{if } m\neq n$$

Then:

Proposition 4.2. The functions $F(z,\tau)=h_n(z;\psi(\tau))$ are eigenfunctions of the transformation (4.40.) with eigenvalue equal to $\lambda_m=K^{\frac{n\alpha}{2}-n+1}$.

Proof. The gauss operator

$$\hat{G}f(z)=(2\pi b)^{-\frac{1}{2}}\int_{-\infty}^{+\infty} f\left(\frac{z}{\sqrt{c}}+x\right)\exp\left(-\frac{x^2}{b}\right)dx$$

transforms an hermite polynomial $h_n(z,a)$ in a hermite polynomial $c^{-\frac{n}{2}}h_n(z;c(a-b))$ if $a>b$

Thus transformation (4.40.) transforms the function $h_n(z;\psi(\tau))$ in thé function $Kc^{-\frac{n}{2}}h_n(z;c(\psi(K\tau)-\gamma(\tau))=$

$$=KK^{-n(2-\alpha)/2}h_n(z;K^{\alpha-2}K^{2-\alpha}\psi(\tau))=K^{\frac{n}{2}\alpha-n+1}h_n(z;\psi(\tau))$$

The proposition is proven.

Remark. Proposition (4.2.) implies that the form

$$H_n(w(t)) = \int_{-\infty}^{+\infty} dz\, h_n\left(\int_{-\pi}^{+\pi} P_\alpha(z,t)\,w(t)\,dt;\ \varphi(z)\right)$$

is eigenfunction of $\partial_K A^*(\alpha)$ with eigenvalue $K^{\frac{n}{2}\alpha - n + 1}$. Further H_n form an orthogonal system of random variables with respect to the gaussian probability measure corresponding to the automodel gaussian stationary random field.

5. The ε -expansion.

It is possible now to give an algorithm which allows to find the effective hamiltonian corresponding to a non gaussian automodel probability distribution. Let us write the effective hamiltonian in this way $H(w) = \sum_i \varepsilon^i f_i(w)$ and expand the corresponding measure in power of ε :

$$\mu_0(dw)\, e^{-\sum_i \varepsilon^i f_i(w)} = \mu_0(dw)\left(1 - \sum_i \varepsilon^i f_i(w) + \frac{1}{2!}\left(\sum_i \varepsilon^i f_i(w)\right)^2 + \dots\right) =$$

5.1.

$$= \mu_0(dw)\left(1 - \varepsilon f_1(w) - \varepsilon^2 f_2(w) + \frac{1}{2}\varepsilon^2 f_1^2(w) + \dots\right)$$

where μ_0 is a gaussian aut.pr.dis.. Let us apply $\partial_K A^*(\alpha)$ to

5.1.:

$$\mu_0(dw)\left(1 - \partial_K A^*(\alpha)\varepsilon f_1 - \varepsilon^2 \partial_K A^*(\alpha) f_2(w) + \frac{1}{2}\varepsilon^2 \partial_K A^*(\alpha) f_1^2(w) \dots\right) =$$

5.2.: $$= \mu_0(dw) \exp \log\left(1 - \varepsilon \partial_K A^*(\alpha) f_1 - \varepsilon^2 \partial_K A^*(\alpha) f_2(w) + \right.$$

$$\left. + \frac{1}{2}\varepsilon^2 \partial_K A^*(\alpha) f_1^2(w)\right) = \mu_0(dw) \exp\left\{-\varepsilon \partial_K A^*(\alpha) f_1 - \varepsilon^2 \partial_K A^*(\alpha) f_2 + \right.$$

$$\left. + \frac{1}{2}\varepsilon^2\left(\partial_K A^*(\alpha) f_1^2 - \left(\partial_K A^*(\alpha) f_1\right)^2\right)\right\}$$

where we have taken only the terms up to the second order in ε because we are going to find an explicit form of the effective hamiltonian only to this degree of accuracy. Thus we have to find $f_1(\omega)$, $f_2(\omega)$ in such a way that:

5.3a. $$\partial_K A^*(\alpha) f_1(\omega) = f_2(\omega)$$

5.3b. $$\partial_K A^*(\alpha) f_2(\omega) - f_2(\omega) = \frac{1}{2}\left(\partial_K A^* f_1^2 - (\partial_K A^*(\alpha) f_1)^2\right)$$

We are studying the effective hamiltonian in a neighbourhood of $\frac{3}{2}$ which is expected to be a bifurcation point and so ε will be defined by $\varepsilon = \alpha - \frac{3}{2}$. If $\alpha = \frac{3}{2}$ the solution of (5.3a.) should be with eigenvalue 1 but since $\alpha \neq \frac{3}{2}$, $\varepsilon = \alpha - \frac{3}{2}$ we can state

5.4. $$\partial_K A^*(\alpha) H_4 = H_4 + (\lambda_4 - 1) H_4 = H_4 + 2\varepsilon \ln K \, H_4$$
$$\lambda_4 = K^{2\varepsilon} \quad , \quad \lambda_4 - 1 = \varepsilon 2 \ln K$$

and so we resolve the equation (5.3a.) up to the accuracy $O(\varepsilon)$, setting

5.5. $$f_1(\omega) = a_1 H_4(\omega)$$

and putting into the equation (5.3b.) the term $O(\varepsilon)$ it will appear as:

5.6. $$(\partial_K A^*(\alpha) - 1) f_2(\omega) = -a_1^2 L(H_4) - 2 \ln K \, a_1 H_4$$

where L is defined by

$$L(H(\omega)) = -\frac{1}{2}\left[\partial_K A^*(\alpha) H^2(\omega) - (\partial_K A^*(\alpha) H(\omega))^2\right]$$

The aim of this section is to find a solution of (5.6.).

The space of forms where we shall look for a solution of (5.6.) is determined by the structure of the operator $L(H(\omega))$. Let us study it in some detail: we use the representation of the preceeding section:

5.7. $$H(\omega) = \int_{-\infty}^{+\infty} d\tau \; F(\pi(\tau), \tau)$$

where $\Pi(\eta)=\int_{-\pi}^{+\pi}dt\,P_\alpha(t,\eta)w(t)$ and $F(x,\eta)$ is some polynomial in x whose coefficients depend on η. We have:

$$\partial_K A^*(\alpha)\, H^2(\omega) = E\Big(\int_{-\infty}^{+\infty}d\eta_1\int_{-\infty}^{+\infty}d\eta_2\, F\Big(\int_{-\pi}^{+\pi}dt_1\, P_\alpha(t_1,\eta_1)K^{-\frac{\alpha}{2}}w'(Kt_1)+$$

5.8. $$+\int_{-\pi}^{+\pi}dt_1\, P_\alpha(t_1,\eta_1)\,x(t_1)\Big)\cdot F\Big(\int_{-\pi}^{+\pi}dt_2\, P_\alpha(t_2,\eta_2)K^{-\frac{\alpha}{2}}w'(Kt_2)+$$

$$+\int_{-\pi}^{+\pi}dt_2\, P_\alpha(t_2,\eta_2)\cdot x(t_2)\,\Big|\,w'\Big)$$

We can proceed in an analogous way as have done for finding the compact form of the eigenfunctions of the differential, the only difference is that now we will make the conditional expectation with respect to the σ -algebra generated by $X_1=\int_{-\pi}^{+\pi}dt_1\, P_\alpha(t_1,\eta_1)\,x(t_1)$ and $X_2=\int_{-\pi}^{+\pi}dt_2\, P_\alpha(t_2,\eta_2)\,x(t_2)$ and we obtain

5.9. $$\partial_K A^*(\alpha)\, H^2(\omega)=\int_{-\infty}^{+\infty}d\eta_1\int_{-\infty}^{+\infty}d\eta_2\,\frac{K^2}{2\pi\sqrt{\det\Gamma_\eta}}\,F\Big(\frac{z_1}{\sqrt{c}}+x_1,K\eta_1\Big)\cdot$$

$$\cdot F\Big(\frac{z_2}{\sqrt{c}}+x_2,K\eta_2\Big)\cdot\exp\Big\{-\frac{(x,\Gamma_\eta^{-1}x)}{2}\Big\}\,dx_1\,dx_2$$

where

$$z_j=\int_{-\pi}^{+\pi}P_\alpha(\eta_j,t)\,w'(t)\,dt\ ,\ c=K^{\alpha-2}\ ,\ x=(x_1,x_2)$$

and Γ_η is a matrix given by

5.10. $$\Gamma_\eta=\begin{pmatrix} E\Big(\Big|\int_{-\pi}^{+\pi}P_\alpha(\eta_1,t)x(t)\,dt\Big|^2\Big) & E\Big(\int_{-\pi}^{+\pi}P_\alpha(\eta_1,t)x(t)\,dt\,\overline{\int_{-\pi}^{+\pi}P_\alpha(\eta_2,t_2)x(t_2)dt_2}\Big) \\ E\Big(\int_{-\pi}^{+\pi}dt_1\int_{-\pi}^{+\pi}dt_2\,\overline{P_\alpha(\eta_1,t_1)}P_\alpha(\eta_2,t_2)\overline{x}(t_1)x(t_2)\Big) & E\Big(\Big|\int_{-\pi}^{+\pi}P_\alpha(\eta_2,t_2)x(t_2)dt_2\Big|^2\Big)\end{pmatrix}$$

making analogous calculations as before we obtain

5.11. $\Gamma_{\tau, ij} = \psi(K\tau_i, K\tau_j) - K^{2-\alpha}\psi(\tau_i, \tau_j) = \gamma(\tau_i, \tau_j)$

where $\psi(\tau_i, \tau_j) = -\sum_{q=-\infty}^{+\infty} \int_{-\infty}^{+\infty} e^{i2\pi q \tau_1} e^{it(\tau_j - \tau_i)} \frac{\rho_0(t)\,\rho_0(2\pi q - t)}{g(t)}\, dt$

Finally we can write the following expression for $L(H(\omega))$:

$$L(H(\omega)) = -\frac{1}{2}\int_{-\infty}^{+\infty} d\tau_1 \int_{-\infty}^{+\infty} d\tau_2 \Big[\frac{K^2}{(\det \Gamma_\tau)^{\frac{1}{2}}} \int_{-\infty}^{+\infty} dx_1 \int_{-\infty}^{+\infty} dx_2\, F\left(\frac{z_1}{\sqrt{c}} + x_1, K\tau_1\right) \cdot$$

$$\cdot F\left(\frac{z_2}{\sqrt{c}} + x_2, K\tau_2\right) \cdot \exp\left(-\frac{(x, \Gamma_\tau^{-1} x)}{2}\right) dx_1 dx_2 - \frac{K^2}{2\pi\gamma(\tau)} \cdot$$

$$\cdot \delta(\tau_1 - \tau_2) \int_{-\infty}^{+\infty} dx_1 \int_{-\infty}^{+\infty} dx_2\, F\left(\frac{z_1}{\sqrt{c}} + x_1, K\tau_1\right) \cdot F\left(\frac{z_2}{\sqrt{c}} + x_2, K\tau_2\right) \cdot$$

$$\cdot \exp\left(-\frac{x_1^2 + x_2^2}{2\gamma(K\tau_1)}\right) \Big]$$

making the integrals with respect to x_1, x_2 and using the theorem of moments of gaussian system of variables (20) (Wick theorem) we obtain a sum only on the connected graphs because the diagonals terms of Γ_τ are equal to $\gamma(K\tau)$. From this representation we can also deduce that we need to consider a space of forms of the type:

5.12. $H(\pi) = \int_{-\infty}^{+\infty} d\tau_1 \int_{-\infty}^{+\infty} d\tau_2\, F(\pi(\tau_1), \tau_1, \pi(\tau_2), \tau_2)$

Repeating the calculations made before we obtain that the eigenfunctions of the differential in the space of forms of the type (5.12.) are the multidimensional hermite polynomials.

We give this statement in more precise terms:

Proposition 5.1. Let $\psi(x,y)$ be defined as in (5.11.) and let $\Gamma_{z,ij}$ be the corresponding matrix. Consider the gaussian random field $\pi(z)$ with dispersion matrix Γ_z :

$$5.13. \quad E\left(e^{i(u_1\pi(z_1)+\cdots+u_n\pi(z_n))}\right) = e^{-\sum_{l,m=1}^{n} u_l u_m \gamma(z_l,z_m)}$$

Let $:\pi^{q_1}(z_1)\cdots\pi^{q_n}(z_n):$ be the Wick polynomial obtained by $\pi^{q_1}(z_1)\cdots\pi^{q_n}(z_n)$ making the ortogonalization with respect to the gaussian distribution (5.13.). Then

$$5.14. \quad \partial_K A^*(d) :\pi^{q_1}(z_1)\cdots\pi^{q_n}(z_n):_\psi = \frac{K^n}{K^{|q|\frac{2-\alpha}{2}}} :\pi^{q_1}(z_1)\cdots\pi^{q_n}(z_n):$$

where $|q| = q_1+\cdots+q_n$

The ortogonalization is just the same as used in euclidean quantum field theory:

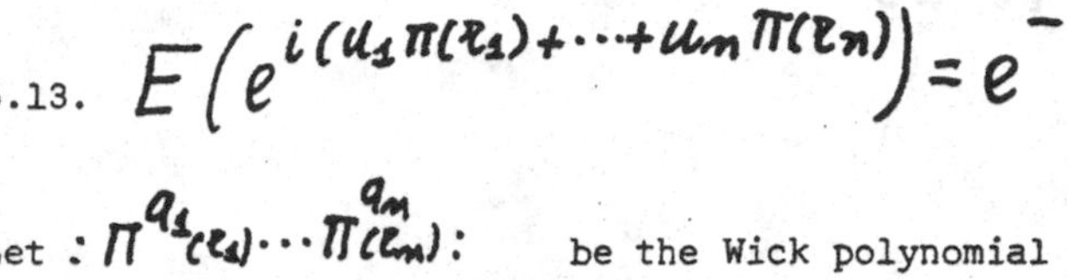

the grafics designed above must be interpreted with the help of the following explanations:

- to every vertex corresponds a point z
- to every line going into the vertex z_i corresponds a random variable $\pi(z_i)$
- the line joining two vertexes corresponds to $\langle\pi(z_1)\overline{\pi(z_2)}\rangle = \gamma(z_1,z_2)$

Then the definition of $:\pi^{q_1}(z_1)\cdots\pi^{q_n}(z_n):_\psi$ is given by

$$:\pi^{q_1}(z_1)\cdots\pi^{q_n}(z_n):_\psi = \pi^{q_1}(z_1)\cdots\pi^{q_n}(z_n) +$$

$$- C_2 \sum_{\{z_i,z_j\}} \langle\pi(z_i)\overline{\pi}(z_j)\rangle\cdot\pi^{q_1}(z_1)\cdots\pi^{q_i-1}(z_i-1)\cdots +$$

$$+ C_4 \sum_{\{z_i z_j z_k z_l\}} \langle\pi(z_i)\overline{\pi}(z_j)\rangle\langle\pi(z_k)\overline{\pi}(z_l)\rangle\pi^{q_1}(z_1)\cdots\pi^{q_i-1}(z_{i-1})\cdots$$

where the sums go over all the possible graphes obtained joining pair of lines of the picture above. We can make use of Proposition (5.1.) for solving equation (5.6.). In fact we can write:

5.14a. $$\delta_1 = a_1 \int_{-\infty}^{+\infty} :\pi^4(z): dz$$

5.14b. $$\delta_2 = \int_{-\infty}^{+\infty}\int_{-\infty}^{+\infty} dz_1 dz_2\, F(\pi(z_1), z_1, \pi(z_2), z_2) =$$

$$= \int_{-\infty}^{+\infty}\int_{-\infty}^{+\infty} dz_1 dz_2 \sum_{\ell_1,\ell_2} F_2^{\ell_1,\ell_2}(z_1,z_2) :\pi^{\ell_1}(z_1)\,\pi^{\ell_2}(z_2):$$

where we have suppressed the index φ but the Wick polynomials must be understood in the sense of Proposition (5.1.). Thus we can write

5.15. $$(\partial_K A^*(\alpha) - 1)\,\delta_2(\omega) = \int_{-\infty}^{+\infty}\int_{-\infty}^{+\infty} dz_1 dz_2 \cdot \sum_{\ell_1,\ell_2} \Big[F_2^{\ell_1,\ell_2}{}_{(Kz_1, Kz_2)} \cdot \frac{K^2}{K^{\frac{\ell_1+\ell_2}{2}(2-\alpha)}} - F_\alpha^{\ell_1,\ell_2}(z_1,z_2) \Big] :\pi^{\ell_1}(z_1)\,\pi^{\ell_2}(z_2):$$

We can also expand the quadratic operator in (5.6.) with the help of hermite polynomials. In fact we have

5.16. $$\delta_1^2 = a_1^2 \int_{-\infty}^{+\infty} dz_1 \int_{-\infty}^{+\infty} dz_2 \,:\pi^4(z_1)::\pi^4(z_2):$$

we can expand δ_1^2 in terms of the eigenfunctions of $\partial_K A^*(\alpha)$

$$:\pi^4(z_1)::\pi^4(z_2): = :\pi^4(z_1)\,\pi^4(z_2): + \sum \langle \pi(z_1)\,\pi(z_2)\rangle \cdot$$

5.17. $$\cdot :\Pi^3(r_1)\Pi^3(r_2): + \sum \langle \Pi(r_1)\Pi(r_2)\rangle^2 :\Pi^2(r_1)\Pi^2(r_2): + \cdots$$

It is possible to see that this sum is equivalent to make the grafic expansion of before and keeping only the connected terms.

Thus it follows from proposition (5.1.) that:

$$\partial_K A^*(d) f_1^2 = a_1^2 \int_{-\infty}^{+\infty} dr_1 \int_{-\infty}^{+\infty} dr_2 \Big[\frac{K^2}{K^{4(2-\alpha)}} :\Pi^4(r_1)\Pi^4(r_2): +$$

$$+ 16 \frac{K^2}{K^{3(2-\alpha)}} \cdot \psi(Kr_1)\psi(Kr_2) :\Pi^3(r_1)\Pi^3(r_2): +$$

5.18. $$+ 72 \frac{K^2}{K^{2(2-\alpha)}} :\Pi^2(r_1)\Pi^2(r_2): \psi^2(Kr_1,Kr_2) +$$

$$+ 96 \frac{K^2}{K^{2-\alpha}} :\Pi(r_1)\Pi(r_2): \psi^3(Kr_1,Kr_2) + 24K^2\psi^4(Kr_1,Kr_2)\Big] dr_1 dr_2$$

and $$\partial_K A^*(d) f_1 = \frac{K}{K^{2(2-\alpha)}} f_1$$

$$(\partial_K A^*(d) f_1)^2 = \frac{K^2}{K^{4(2-\alpha)}} f_1^2 = \int_{-\infty}^{+\infty} dr_1 dr_2 \frac{K^2 a_1^2}{K^{4(2-\alpha)}} [:\Pi^4(r_1)\Pi^4(r_2): +$$

5.19. $$+16\psi(r_1,r_2) :\Pi^3(r_1)\Pi^3(r_2): + 72 :\Pi^2(r_1)\Pi^2(r_2): \psi^2(Kr_1,Kr_2) +$$

$$+ 96 :\Pi(r_1)\Pi(r_2): \psi^3(Kr_1,Kr_2) + 24K^2\psi^4(Kr_1,Kr_2)]$$

The integral $\int \psi^4(Kr_1,Kr_2)dr_1 dr_2$ diverges because of the periodic property of ψ : $\psi(r_1+1,r_2+1) = \psi(r_1,r_2)$, but it gives a constant which will give no contribution to the probability distribution. This divergence arises because we calculate directly the effective hamiltonian which is a sum of some potential over all the lattice points: the same calculation for the potential would have given a finite term.

Putting (5.18.), (5.19.), (5.15.) into eq. (5.6.) and neglecting the term $2\ln K\, a_1 H_4$ we obtain the following equations:

$$F_2^{4,4} = 0$$

5.20.

$$F_2^{(3,3)}(Kr_1, Kr_2)K^{2-3(2-\alpha)} - F_2^{(3,3)}(r_1, r_2) =$$

$$= 16a_1^2\left(K^{2-\alpha}\psi(Kr_1, Kr_2) - \psi(r_1, r_2)\right)K^{4\alpha-6} =$$

$$= 16a_1^2\left(K^{2-3(2-\alpha)}\psi(Kr_1, Kr_2) - \psi(r_1, r_2)\right) + O(\varepsilon)$$

putting $O(\varepsilon)$ in the higher order equations we obtain that

$$F_2^{(3,3)}(r_1, r_2) = 16\psi(r_1, r_2)a_1^2$$

5.21. $$F_2^{(2,2)}(r_1, r_2) = 72\,\psi^2(r_1, r_2)\,a_1^2$$

$$F_2^{(1,1)}(r_1, r_2) = 96a_1^2\,\psi^2(r_1, r_2)$$

Proposition 5.2.

The function $\psi(r_1, r_2)$ belongs to $C^1(\mathbb{R}^2)$ is real, periodic and symmetric and has the following asymptotic

$$\psi(r_1, r_2) \sim |r_1 - r_2|^{\alpha-2}, \quad |r_1 - r_2| \longrightarrow \infty$$

Proof. The proposition follows immediately from the definition (5.11.) of $\psi(r_1, r_2)$. In fact the Fourier transform of $\psi(r_1, r_2)$ is given by

5.22. $$\psi(u_1, u_2) = -\sum_{K=-\infty}^{+\infty} \frac{\rho_0(u_1)\rho_0(u_2)}{g(u_1)}\,\delta_p(u_1 + u_2 + 2\pi K)$$

It is sufficient to look at the behaviour of $\psi(u_1, u_2)$ in the origin

$$\psi(u_1, u_2) \sim \delta(u_1 + u_2)\,\frac{u_1}{|u_1|^{1+\alpha}} \cdot \frac{u_2}{|u_2|^{1+\alpha}} \cdot |u_1|^{1+\alpha}$$

From which it follows that:

$$\psi(r_1,r_2) \sim |r_1-r_2|^{-(1-\alpha)-1} = \frac{1}{|r_1-r_2|^{2-\alpha}}$$

From the definition of δ_2 (5.14b.), it is clear that we need $F_2^{\ell_1,\ell_2}(r_1,r_2)$ to be integrable on $\mathbb{R}^2$ and to have singularities which are integrable on the plane.

First we note that we caw subtract from $F_2^{(3,3)}$, $F_2^{(2,2)}$ its behaviour at large $|r_1-r_2|$ and the equations of the type (5.20.) will be satisfied just the same. In fact the right hand side of (5.20.) will not change.

5.23.
$$K^{4\alpha-6}\,16\left(K^{2-\alpha}\psi(Kr_1,Kr_2)-\psi(r_1,r_2)\right) =$$
$$=K^{4\alpha-6}\cdot 16\left[K^{2-\alpha}\left(\psi(Kr_1,Kr_2)-|Kr_1-Kr_2|^{\alpha-2}\right)-\left(\psi(r_1,r_2)-|r_1-r_2|^{\alpha-2}\right)\right]$$

we need to subtract the behaviour at large $|r_1-r_2|$ only for $F_2^{(3,3)}, F_2^{(2,2)}$ because $F_2^{(1,1)}$ has a sufficiently good decrease property at infinite.

But we have that $F_2^{(2,2)}$ now has a non-integrable singularity for r_1, r_2 as a consequence of the subtraction. So we shall set

5.24. $$F_2^{(2,2)}=72\left(\psi^2(r_1,r_2)-\mathrm{Reg}\left\{|r_1-r_2|^{2(\alpha-2)}\right\}\right)a_1^2$$

where $\mathrm{Reg}\left\{\frac{1}{|x|^{4-2\alpha}}\right\}$ is a generalized function defined by

$$\left(\mathrm{Reg}\left\{\frac{1}{|x|^{4-2\alpha}}\right\},\varphi(x)\right)=\int_{-\infty}^{+\infty}\frac{1}{|x|^{4-2\alpha}}\left(\varphi(x)-\varphi(0)\chi_{[-1,+1]}(x)\right)dx$$

Now it is possible to show the following identity between generalized functions

5.25. $$\mathrm{Reg}\left\{\frac{1}{|Kr_1-Kr_2|^{4-2\alpha}}\right\}=K^{2\alpha-4}\left[\mathrm{Reg}\left\{|r_1-r_2|^{2\alpha-4}\right\}+g\,\delta(r_1-r_2)\right]$$

Which follows from the definition:

$$5.26.\quad \left(\mathrm{Reg}\left\{\frac{1}{|Kx|^{4-2d}}\right\}, \varphi(x)\right) \overset{\text{def.}}{=} \left(\mathrm{Reg}\left\{\frac{1}{|x|^{4-2d}}, \varphi\left(\frac{x}{K}\right)\right)K^{-1}\right.$$

Thus

$$\left(\mathrm{Reg}\frac{1}{|Kx|^{4-2d}}, \varphi(x)\right) = K^{-1}\int_{-\infty}^{+\infty} dx \frac{1}{|x|^{4-2d}}\left(\varphi\left(\frac{x}{K}\right) - \chi_{[-1,+1]}(x)\right) =$$

$$\frac{x}{K} \to x_0 \quad = \int_{-\infty}^{+\infty} \frac{1}{|Kx_0|^{4-2d}}\left(\varphi(x_0) - \varphi(0)\chi_{[-1,+1]}(Kx_0)\right) dx_0 =$$

$$= K^{2d-4}\left(\mathrm{Reg}\left\{\frac{1}{|x_0|^{4-2d}}\right\}, \varphi(x_0)\right) + 2K^{2d-4}\int_{K^{-1}}^{1} \frac{\varphi(0)}{|x_0|^{4-2d}} dx_0\Big) =$$

$$5.27.\quad = K^{2d-4}\left(\mathrm{Reg}\left\{\frac{1}{|x_0|^{4-2d}}\right\}, \varphi(x_0)\right) +$$

$$+ 2\,\frac{K^{2d-4}}{2d-3}\left(1 - K^{3-2d}\right)\left(\delta(x_0), \varphi(x_0)\right)$$

From (5.24.) and the fact that $\frac{1-K^{3-2d}}{2d-3} \sim 2\ln K$ we obtain that the term neglected in equation (5.20.) is equal to the contribute given by the -function to the left hand side of (5.6.) if

$$a_1^2 K^{4d-6} g = a_1, \quad a_1 = \frac{1}{K^{4d-6}}\frac{1}{g}$$

and so the ε -expansion is completely resolved up to the second order. Now we are going to interpret these results in term of the effective hamiltonian for the field $\{\xi_n\}_{-\infty}^{+\infty}$.

We have found that the non gaussian term of the hamiltonian has the following form

5.28.
$$f_2(\omega)=\int_{-\infty}^{+\infty}dz_1\int_{-\infty}^{+\infty}dz_2\, a_1^2\Big(16\left[\Psi(z_1,z_2)-|z_1-z_2|^{\alpha-2}\right]\cdot$$
$$\cdot\, :\Pi^3(z_1)\Pi^3(z_2):+72\left[\Psi^2(z_1,z_2)-\mathrm{Reg}\,|z_1-z_2|^{2\alpha-4}\right]\cdot$$
$$:\Pi^2(z_1)\Pi^2(z_2):\;+\;96\,\Psi^3(z_1,z_2):\Pi(z_1)\Pi(z_2):\Big)$$

Let us rewrite (5.28.) in terms of $\eta(t)$.

We observe that

5.29.
$$\Pi(z)=\int_{-\pi}^{+\pi}P_\alpha(z,t)\,w(t)\,dt=\sum_{j=-\infty}^{+\infty}\int_{-\pi}^{+\pi}g(t+2\pi j)\,e^{iz(t+2\pi j)}\,w(t+2\pi j)\,dt=$$
$$=\int_{-\infty}^{+\infty}g(t)\,e^{izt}\,w(t)\,dt=\int_{-\infty}^{+\infty}g(t)\,e^{izt}\,g^{-1}(t)\,(e^{-it}-1)^{-1}\,\eta(t)\,dt=$$
$$=\int_{-\infty}^{+\infty}e^{izt}\,S(t)\,\eta(t)\,dt$$

where $S(0)\neq 0$. Since we have already subtracted from the first two terms of (5.28.) the gaussian behaviour we compare only the third term of (5.28.) with the gaussian term.

We will write the last term of (5.28.) in the following useful form:

5.30.
$$\int_{-\infty}^{+\infty}dz_1\int_{-\infty}^{+\infty}dz_2\int_{-\infty}^{+\infty}dt_1\, e^{iz_1t_1}S(t_1)\,\eta(t_1)\int_{-\infty}^{+\infty}dt_2\, e^{iz_2t_2}S(t_2)\cdot$$
$$\cdot\,\eta(t_2)\,\Psi(z_1,z_2)=\int_{-\infty}^{+\infty}\int_{-\infty}^{+\infty}dt_1\,dt_2\,\tilde{\Psi}^{(3)}(t_1,t_2)\,S(t_1)\,S(t_2)\,\eta(t_1)\,\eta(t_2)=$$
$$=\int_{-\pi}^{+\pi}dt_1\int_{-\pi}^{+\pi}dt_2\,\eta(t_1)\,\eta(t_2)\Big(\sum_{j_1,j_2=-\infty}^{+\infty}\tilde{\Psi}^3(t_1+2\pi j_1,\,t_2+2\pi j_2)\Big)S(t_1)\,S(t_2)$$

we want to compare the contribution to the effective interaction between two spins ξ_{p}, ξ_{p+n} due to the non gaussian term with the one due to the gaussian term. We write the last one in the following way:

5.31. $$\iint |e^{-it}-1|^{-2} \mathcal{G}^{-1}(t_1,t_2)\delta(t_1+t_2)\eta(t_1)\eta(t_2)\,dt_1 dt_2$$

For this aim we have to study the analiticity properties of $\psi(z_1,z_2)$ for all the values of $|z_1-z_2|$. Thus we have to study the analiticity properties of the series (5.22.).

Let us study before the term in (5.22.) with $u_1+u_2=0$

5.32. $$\begin{aligned}\psi(u_1,u_2) &= \frac{u_1^2}{|u_1|^{2d+2}}\left(|u_1|^{-d-1}+\mathcal{G}_0(u_1)\right)^{-1} = \\ &= |u_1|^{-d+1}\left(1-|u_1|^{d+1}\mathcal{G}_0(u_1)+|u_1|^{2d+2}\mathcal{G}_0^2(u_1)\cdots\right)\end{aligned}$$

where $\mathcal{G}_0(u_1)$ is an analytical function of u_1, and we have developed the factor $(1+|u_1|^{d+1}\mathcal{G}(u_1))^{-1}$ in Taylor series in the neighbourhood of the origin, thus we can write

$$\psi(u_1,u_2) = |u_1|^{-d+1} + u_1^2\mathcal{G}_0(u_1) + |u_1|^{d+3}\mathcal{G}_0^2(u_1) + \cdots$$

The second term gives no singularity in the origin because it is analytical while the first and the third will give

5.33. $$\psi(z_1,z_2) \sim \frac{1}{|z_1-z_2|^{2-d}} + \frac{1}{|z_1-z_2|^{d+4}}$$

The singularity arising by terms where $u_1+u_2=2\pi K$ gives a contribution no larger than

5.34. $$\frac{1}{|z_1-z_2|^{d+2}}$$

as it is possible to ses from elementary considerations.

Thus the main contribution to the effective interaction potential comes from the first term in (5.33.) and we have that

5.35. $$\tilde{\Psi}^3(t_1,t_2) = |t|^{-[3(\alpha-2)]-1}\delta(t_1+t_2) = |t|^{5-3\alpha}\delta(t_1+t_2)$$

which must be compared with the behaviour of the gaussian term

5.36. $$\frac{1}{t^2}|t|^{1+\alpha} = |t|^{\alpha-1}$$

(5.36.) given a potential which decays as $n^{-\alpha}$, while (5.35.) gives $n^{3\alpha-6}$ which is slower than $n^{-\alpha}$. Further investigation can be made in two possible directions:

A) To study the higher order equations of the ε -expansion

B) regularize also the third term of (5.28.). In this case we would obtain the correct behaviour of the interaction potential and also the unicity of the solution of eq. (5.6.) and so B seems to be the most fruitful way to investigate.

6. Some new results for spin systems *)

Let us consider a d-dimensional lattice system, let $b(r)$ be the correlation function of a Gaussian automodel random field

6.1. $$b(r) = E\,\xi_{x+r}\,\xi_x = \int e^{2\pi i(\lambda,x)}\rho(\lambda)d\lambda$$

where $\lambda \doteq (\lambda_1,\dots,\lambda_d)$, $x+r, x\in Z^d$, $\rho(\lambda) = \rho(\lambda_1,\dots,\lambda_d)$

Then the Gaussian aut.pr.distr. can be written as a Gibbs distribution in the

---

*) Recently E.I.Dinaburg and Ja.G.Sinai proved that for $1<\alpha<\frac{3}{2}$ the gaussian automodel distribution appears as the limit probability distribution at $\beta=\beta_{cr}$ for some one-dimensional translationally invariant system with the long range potential $U(r)\sim\frac{c}{r^\alpha}$

form:

6.2. $$e^{-\frac{1}{2}\sum_{t_1,t_2\in Z^d} U(t_1-t_2)\xi_{t_1}\xi_{t_2}}$$

where $U(t) = \int e^{2\pi i(\lambda,x)} \rho^{-1}(\lambda)d\lambda$

It is possible to find, for $d>2$ a case when $\rho^{-1}(\lambda)$ is a real analytical function and so $U(t)$ is a function exponentially decreasing

$$U(t) \leq Const. \cdot e^{-A|t|}$$

It is probable that, for $\alpha<\frac{3}{2}$ such a probability distribution appears as limit distribution for $\beta=\beta_{cr}$ for a large class of spin systems with short range interaction potential.

7. Automodel probability distributions in the continuous case*)

7.1. Gaussian case.

D.7.1. Automodel probability distributions.

Let P be a probability measure defined on the space of generalized functions $f(t)$ on R^ν. Consider the following transformation:

7.1. $$f(t) \to \lambda^{-\kappa} f(\lambda t) \quad \lambda\in[0,\infty)$$

If $f(t)$ and $\lambda^{-\kappa}f(\lambda t)$ have the same probability distributions, then P

---

*) In this section are given some results obtained by R.L.Dobrushin. These results are collected in a paper which must be published.

is an automodel probability distribution with parameter K. It is possible to establish the following connection with the discrete case.

Suppose that $\mathcal{S}$ is the space of test functions which must satisfy smoothness properties. Consider the space of functions M which contains $\mathcal{S}$ and the characteristic functions of the unit squares. If it is possible to extend the integral $\int \xi(t)\varphi(t)\,dt$ to the space M then we can define the fol-fowing discrete random field: for each $u \in Z^\nu$

7.2. $$\xi_u = \int_{\Lambda_u} \xi(t)\,dt$$

where Λ_u is the unit square with center in the point u

Proposition 7.1. If $\xi(t),\ t \in R^\nu$ is automodel in the sense of D.7.1. with parameter K then $\{\xi_u, u \in Z^\nu\}$ is automodel with respect to the Ka-danoff's renormalization group with $\alpha = 2 - \frac{2K}{\nu}$ Gaussian automodel gen. random field.

A Gaussian generalized r.f. is defined by means of its spectral measure

7.3. $$B(\varphi,\psi) = E\left(\int \xi(t)\varphi(t)dt \cdot \int \xi(t)\psi(t)dt\right) = \int \tilde{\varphi}(k)\tilde{\psi}(k)\,G(dk)$$

where $G(dk)$ is a measure on R^ν. Let us introduce the polar coordinates: Q_ν is the unit sphere, $\underline{e}$ the unit vector, $d \in [0,\infty)$ then $(x_1,\dots,x_\nu) \to (d,\underline{e})$

Then:

Proposition 7.2. A generalized gaussian random field is automodel when its spectral measure is given by

7.4. $$G(d \in C, \underline{e} \in E) = \bar{G}(E) \int_C d^{2K-1}\,d\alpha$$

where $\bar{G}$ is any measure on the surface of the unit sphere Q_ν.

It is possible to construct aut. discrete gauss. random field from this generalized gauss. field. This can be done when the measure $\bar{G}$ has density $g(e),\ e \in Q_\nu$ such that there exist constants $c, C > 0$ for which

7.5. $$0 < c < g(e) < \nu$$

then it is possible to construct a discret Gauss. ran.field if $0<k<\nu$.
A very important example is the isotropic gauss. ran. f. for which

$$g(e) = c\,|e|^{2k-\nu}$$

After the discretization it is obtained a class of discret Gaussian fields which contains that of Sinai. The two class coincide only when $\mathcal{G}$ has a density.

It is important to emphasize that in this kind of approach it is considered a special class of gen. random field. This is determined choosing the test function to belong to $\mathcal{S}_2$ where $\mathcal{S}_2$ is the space of smooth functions such that if $\varphi \in \mathcal{S}_2$ then:

7.6. $$\int \varphi(x)\, x_1^{j_1} x_2^{j_2} \cdots x_\nu^{j_\nu}\, dx_1 \cdots dx_\nu = 0$$

with $j_1 + \cdots + j_\nu \leq 2$. Such a process is said to be defined "over $\mathcal{S}_2$".
It can be shown that the condition introduced on the test functions implies that we consider random fields with independent increments (22). The use of such fields is useful because it is possible to allow more critical behaviour in the origin for $\mathcal{G}(d\epsilon G, \underline{e} \in E)$ in order it to be gaussian.
In this case in fact the following condition must be satisfied:

7.7. $$\int_{|x|\leq 1} \mathcal{G}(dx)\, |x|^{22} < \infty$$

Using the same concept the following results can be obtained: let $:\varphi^n:_2$ be the Wick polynomial made with respect a gauss.gen. field then in the limit $\lambda \to \infty$ in (7.1.) $:\varphi^n:_2$ converges to the massless gaussian field.

7.2. Non gaussian case

Let $\xi(t)$ be a gen. rand. f. and P the corresponding state, that is the probability measure defined on the space of generalized functions $\mathcal{S}'$.

Let us consider a class of functionals defined on $\mathcal{S}'$, $\{\Phi_\varphi, \varphi \in \mathcal{S}'\}$ and suppose that they belong to $L^2(P)$. We can define in $L^2(P)$ and also in this class of functionals a shift transformation induced by the shift transformation of $\xi(t)$

7.8. $$\Phi_{\varphi(x-a)} = E_a \Phi_\varphi$$

Then we say that the generalized r.proc. $\zeta(x)$ is subordinated to $\xi(t)$ or induced by $\xi(t)$ if ζ is given by

7.9. $$\zeta(x) = \psi(\xi(x+a)),\ x \in R^\nu$$

It can be shown that the class of gen.r.pr. induced by a gen. Gauss. r.f. can be represented using the Ito integral. Let $Z(d\lambda)$ be the spectral normal random measure corresponding to $\xi(t)$

7.10. $$\xi(t) = \int e^{it\lambda} Z(d\lambda)$$

Then it is possible to show that all the generalized random processes of the type (7.9.) can be written as:

7.11. $$\Phi_\varphi = \sum_{n=1}^{\infty} \frac{1}{n!} \int \hat{\varphi}_n (x_1 + \cdots + x_n) \hat{h}_n (x_1, \ldots, x_n) Z(dx_1) \ldots Z(dx_n)$$

if $\hat{h}_n(x_1, \ldots, x_n)$ is such that

$$\sum_n \frac{1}{n!} \int |\hat{h}_n (x_1, \ldots, x_\nu)|^2 |x_1 + \cdots + x_n|^{-q} G(dx_1) \ldots G(dx_n) < \infty$$

Let us suppose now that the gauss.gen.ran. is automodel in the sens of(D.7.1.) with parameter K_0

Then

Proposition 7.3. Φ_φ is automodel (non gaussian) generalized ran. field with parameter K if

7.12. $$\hat{h}_n(\lambda x_1,\dots,\lambda x_n) = \lambda^{-n\kappa_0+\kappa}\,\hat{h}_n(x_1,\dots,x_n)$$

It is also possible to represent $\hat{h}_n$ for $0<\kappa<\frac{\nu}{2}$ when the Gaussian automodel prob. distr. has parameter

7.13. $$\hat{h}_n(\lambda x_1,\dots,\lambda x_n) = \prod_{j=1}^{n} |k_j|^{-\kappa_0+\frac{\kappa}{n}}\, g_n\left(\frac{k_j}{|k_j|}\right)$$

where the g_n are bounded functions.

It is possible to formulate Proposition (7.3.) in another interesting way. Let us evaluate the wick polynomial $:\varphi(x)^M:_g$ with respect to the gaussian automodel g then:

Proposition 7.4. $\phi(x) = :\varphi(x)^M:_g$ is a non gaussian automodel gen.ran.f.

We note for example that discretizing $:\varphi^2(x):_g,\ \nu=1$ one obtains a Rosenblatt process (see G.Gallavotti, G.Iona-Lasinio (23)).

All the fields represented by (7.13.) can be discretized and so one obtaines non gaussian automodel pr.distr.. There is also a new branching aut-pr. distr.:the white noise.

This result is new with respect to those obtained by Bleher e Sinai and it is not clear if this new bifurcating point can be found as Gibbs distribution for $\beta=\beta_{cr}$ for some models.

Thus we can draw the following picture

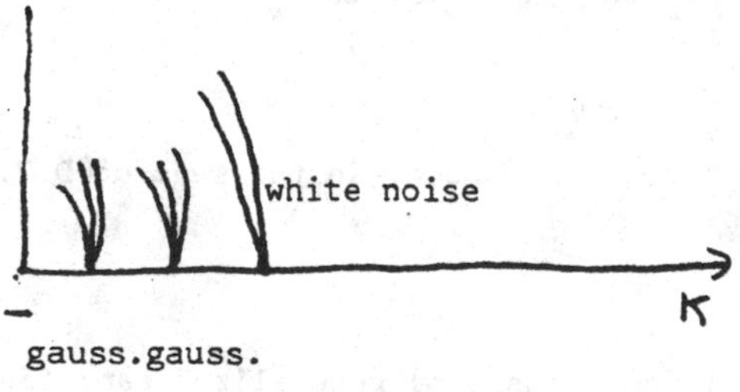

K: parameter of automodelity.

The curves indicate a family of different non gauss automodel prob.distr.

as function of K For every aut. gauss. pr. distr. it is possible to find the branching of aut. pr. distr. as before.

Acknowledgment. I am very grateful to professors P.M.Bleher, Ja.G.Sinai, R.L. Dobrushin for their useful explanations and for very interesting discussions with them.
The author acknonledeges very Kind hospitality of the Chair of Probability theory of the Moscow State University, Lomonosov.

Bibliography

Section 1.

1) Gnedenko, B.V., Kolmogorov, A.N., Limit distributions for sums of independent random variables, Gostechisdat, Moscow (1949).

2) Ibragimov, I.A., Linnik, Ju, V:
Independent and stationary sequences of random variables, Nauka, Moscow, (1965), English Transl. Noordhoff, Groningen, (1971).

3) Gnedenko, B.V., Course on Probability Theory, Nauka, Moscow (1965).

4) Dobrushin, R.L., Funz. An. 2, 4, 31, (1968).

5) Lanford O.E., Ruelle D. Comm. Math. Phys. 13, 194, (1969).

6) Nakhapitan, B.C., Dokl. Armianskaia Ak. Nauk. V61, (1975).

7) Malishev, V.A.,: Dokl. Ak. Nauk. U.R.S.S. 224,(1975).

8) Dobrushin R.L., Tirozzi, B.: to appear on Comm. Math. Phys.

9) Gallavotti, G.,: Martin Lõf, A. Nuovo Cimento, 25B, 1, 425, (1975).

10) Dobrushin, R.L., Mat. Sbornik. 94, 136, (1974).

11) Kadanoff, L. et al. Rev. Mod. Phys. 39, 395, (1967).

12) Ja. G. Sinai, Teor. Ver. i ee Prim., 21, 1, (1976).

13) Arnold, V.I.,: Lectures on bifurcation and versal families, Usp. Mat. Nauk 27, 5, 119, (1972).

14) Kosterlitz, J.M.,: Critical properties of the one dimensional Ising model with long range interactions, University of Birmingham preprint.

15) P.M. Bleher, Ya.G.Sinai, Comm. Math. Phys. 45, 247, (1975).

16) Arnold, V.I., Matematiceskie metodi classiceski mekaniki, Nauka, Moscow, (1974).

17) Dobrushin, R.L., Teor. Ver. i ee Prim. 15, 3, 469 (1970).

18) Ito, K., Japan Journal of Mathematics, 22, 63, (1949).

19) Ja. G. Sinai, Introduction to ergodic theory, Edited by Erevan and Moscow University, (1973).

20) Ghikhman, I. I.,Skhorokhod,A.V., Theory of random processes, Ed. Nauka, Moscow 1973, Transl. Springer-Verlag, Berlin (1975).

21) Bleher, P.M. ε-expansion in Kadanoff renormalization group, Preprint.

22) Iaglom, A.M., Teor. Ver. i ee Prim., Vol. 2. N.3, p. 292, (1957).

23) G., Gallavotti, G.Iona-Lasinio, Comm. Math. Phys. 41, N.3, 301, (1975).

CENTRO INTERNAZIONALE MATEMATICO ESTIVO
(C.I.M.E.)

BASIC PROPERTIES OF ENTROPY IN QUANTUM MECHANICS

A. WEHRL

Institut für Theoretische Physik
Universität Wien, Austria

BASIC PROPERTIES OF ENTROPY IN QUANTUM MECHANICS

ALFRED WEHRL
Institut für Theoretische Physik
Universität Wien
Austria

Introduction

Entropy is one of the most important quantities in physics, for it governs the behaviour of macroscopic systems and relates macroscopic and microscopic quantities. In this lecture, I do not want to go into the thermodynamic foundations of entropy but shall rather concentrate on some mathematical aspects of it.

From the mathematical point of view, entropy can be considered as a measure of the intrinsic dispersion (degree of mixing, impurity, uncertainty, lack of information, amount of chaos) of a quantum state. (However, one should not forget about the fact that, at least in equilibrium, entropy is a measurable quantity). A state in ordinary quantum mechanics being described by a density matrix ρ, i.e. a linear operator, ≥ 0, with trace = 1, its entropy is given by v. Neumann's formula[1)]

$$S(\rho) = -k_B \operatorname{Tr} \rho \ln \rho.$$

In what follows we shall always put Boltzmann's constant $k_B = 1$.

V. Neumann's formula resembles very much Shannon's formula for the information of a finite, classical probability distribution[2)]: if $p = (p_1, \ldots, p_n)$ $(p_i \geq 0,$

$\sum p_i = 1$) is such a distribution, then $I(p) = -\sum p_i \ln p_i$. In fact, many of the theorems on quantum-mechanical entropy are more or less obvious generalizations of results known from classical information theory. However, in some instances, the proofs are rather difficult (for instance, for the strong subadditivity), and, furthermore, there are classical results that are false in the quantum case.

Simple Properties

(a) Domain and range of entropy. Since a density matrix, being a compact operator, can be diagonalized,

$$\rho = \sum \rho^{(i)} P_i$$

(the P_i being one-dimensional projections), with $\rho^{(i)} \geq 0$, $\sum \rho^{(i)} = 1$,

$$S(\rho) = \sum s(\rho^{(i)})$$

($s(x) := -x \ln x$ if $x > 0$, $= 0$ if $x = 0$); $s(\rho^{(i)})$ is always ≥ 0, so that $S(\rho)$ is always defined, ≥ 0, but possibly $= \infty$, if the Hilbert space is infinite-dimensional. One checks easily that the range of S is the whole extended real half-line $[0,\infty]$.

Because $s(x) = 0 \rightarrow x = 0$ or 1, $S(\rho) = 0$ implies ρ = pure, i.e. a one-dimensional projection. Pure states are determined by a vector ψ ("wave function"): since one-dimensional projections are of the form $\rho = |\psi)(\psi|$, in this case

$$\mathrm{Tr}\, \rho A = (\psi | A\psi).$$

Since the wave-function, according to the principles of quantum mechanics, contains the maximal information that can be obtained by measurements, it is plausible that exactly in that case the entropy must be zero.

(b) Partial isometric invariance[3]. $S(\rho)$ depends on the positive eigenvalues of ρ only, therefore, any two density matrices with the same positive eigenvalues (with the same multiplicities) have the same entropy. Thus, if U is an arbitrary unitary operator, then $S(\rho) = S(U \rho U^*)$. Or, more generally, let H_1, H_2 be two Hilbert spaces, ρ_1, ρ_2 be two density matrices in H_1, or H_2, resp., $V: H_1 \to H_2$ be a partial isometry with initial domain $\mathcal{D}_1$ and final domain $\mathcal{D}_2$ such that Ran $\rho_i \subset \mathcal{D}_i$ $(i = 1,2)$ and $\rho_2 = V \rho_1 V^*$, then $S(\rho_1) = S(\rho_2)$.

(c) Monotonicity with respect to mixing. Call two density matrices ρ_1, ρ_2 equivalent ($\rho_1 \sim \rho_2$), if they have the same positive spectrum. If $\rho = \lambda_1\rho_1 + \lambda_2\rho_2$ ($\lambda_i \geq 0$, $\lambda_1 + \lambda_2 = 1$), then $S(\rho) \geq S(\rho_1) = S(\rho_2)$. Heuristically speaking: mixing of equivalent states increases entropy.
Proof: see section "Inequalities", concavity. Note, however, that concavity does not follow from monotonicity.

Uhlmann Theory[4,5]

If one does not take physics into account, there are of course many other measures of intrinsic dispersion besides entropy. The most general concept is due to Uhlmann and states that whatever quantity is introduced as such a measure, it should fulfil properties (b) and (c) from last section. Uhlmann defines a density matrix ρ to be "more mixed", or "more chaotic", than another one, say ρ' (and writes $\rho \succ \rho'$, or $\rho' \prec \rho$), if ρ is in the (weakly) closed convex hull of the set of density matrices that are equivalent (in the sense of (b)) to ρ'.
Let us shortly state some results.

Main theorem. Let $\rho^{(1)}, \rho^{(2)}, \ldots$ (or $\rho'^{(1)}, \rho'^{(2)}, \ldots$, resp.) be the eigenvalues of ρ (or ρ', resp.), arranged in decreasing order and repeated according to multiplicity. $\rho \succ \rho'$ if, and only if, $\rho^{(1)} \leq \rho'^{(1)}, \rho^{(1)} + \rho^{(2)} \leq \rho'^{(1)} + \rho'^{(2)}, \ldots, \rho^{(1)} + \rho^{(2)} + \ldots + \rho^{(n)} \leq \rho'^{(1)} + \rho'^{(2)} + \ldots + \rho'^{(n)}, \ldots$

Order relations. From this theorem it follows immediately that the relation is transitive, and that $\rho \succ \rho'$, $\rho' \succ \rho$ implies $\rho \sim \rho'$, so that $\succ$ is a preorder relation.

Lattice structure of density matrices. One can show that, with respect to $\succ$, the equivalence classes of density matrices form a lattice. There exists always a "smallest" element, namely, the pure states, but only if the Hilbert space is finite-dimensional, a "biggest one", namely $(\dim H)^{-1}\cdot 1$.

Convex and concave functions. $\rho \succ \rho'$, if, and only if, for every convex (or concave) function f, ≥ 0, with $f(0) = 0$, $\mathrm{Tr}\, f(\rho) \leq \mathrm{Tr}\, f(\rho')$ (or ≥ 0, resp.). In particular, $\rho \succ \rho' \rightarrow S(\rho) \geq S(\rho')$, but the converse is not true.

Coarse-graining. Let P_i be a family of pair-wise orthogonal projections with $\sum P_i = 1$. Then $\rho \prec \sum P_i \rho P_i$. (In theories about the measurement process, this is sometimes called "reduction of a state"). If, in addition, there exists a "coarse-grained" density matrix $\rho_c = \sum \lambda_i P_i$ such that $\mathrm{Tr}\, \lambda_i P_i = \mathrm{Tr}\, \rho P_i$, then $\sum P_i \rho P_i \prec \rho_c$.

α-Entropies

Among the measures that are compatible with Uhlmann's order relation, the quantum analogues of Renyi's entropies[2] play a distinguished role. Let

$$S_\alpha(\rho) = \frac{1}{1-\alpha} \ln \mathrm{Tr}\, \rho^\alpha$$

for $\alpha > 0, \neq 1$; $S_0(\rho) = \ln \dim \mathrm{Ran}\, \rho$ (the quantum analogue of the Hartley entropy), $S_1(\rho) = S(\rho)$, $S_\infty(\rho) = -\ln \|\rho\|$. Then $S_\alpha(\rho)$ is decreasing in α[6], finite for $\alpha > 1$, continuous with respect to the trace norm for $\alpha > 1$ since

$$\left|(\mathrm{Tr}\, \rho^\alpha)^{1/\alpha} - (\mathrm{Tr}\, \rho'^\alpha)^{1/\alpha}\right| \leq (\mathrm{Tr}\, |\rho - \rho'|)^{1/\alpha}$$

(this is a consequence of the triangle inequality[7] for the v.Neumann-Schatten classes). Also

$$S(\rho) = -\frac{d}{d\alpha} \mathrm{Tr}\, \rho^\alpha \Big|_{\alpha=1}.$$

α-entropies look very much like the right entropy, in particular, they are additive (see section "Inequalities (Two Spaces)") and have been used on several occasions, e.g. in non-equilibrium statistical mechanics.

Continuity Properties of Entropy

If the Hilbert space is finite-dimensional, entropy is clearly continuous. In the infinite-dimensional case, entropy is discontinuous with respect to the trace norm, because every "ball" $\{\rho': \mathrm{Tr}\,|\rho - \rho'| < \varepsilon\ (\varepsilon > 0)\}$ contains density matrices with infinite entropy. This can be shown explicitly as follows: let $\rho^{(1)} \geq \rho^{(2)} \geq \ldots$ be the eigenvalues of ρ. Choose N such that

$$\sum_{i=N}^{\infty} \rho^{(i)} < \varepsilon.$$

Let ρ' have the same eigenvectors as ρ, but eigenvalues $\rho'^{(i)} = \rho^{(i)}$ for $i < N$,

$$\rho'^{(i)} = \frac{C}{i(\ln i)^2}$$

for $i \geq N$, provided that

$$\sum_{N}^{\infty} \rho^{(i)} > 0,$$

otherwise one can assume that $\rho^{(N-1)} > 0$, then let $\rho'^{(i)} = \rho^{(i)}$ for $i < N - 1$, $\rho'^{(N-1)} = \rho^{(N-1)} - \varepsilon'$ $(\varepsilon' < \varepsilon)$, $\rho'^{(i)}$ for $i \geq N$ as above. In both cases, C is to be chosen such that

$$\sum_{i=1}^{\infty} \rho'^{(i)} = 1.$$

Then, $\mathrm{Tr}\,|\rho - \rho'| < \varepsilon$, but $S(\rho') = \infty$.

<u>Lower semi-continuity for entropy</u>. Since $S_\alpha(\rho)$ is continuous for $\alpha > 1$, and $S(\rho) = \lim_{\alpha\to 1} S_\alpha(\rho) = \sup_{\alpha>1} S_\alpha(\rho)$, $S(\rho)$ is lower semi-continuous. Therefore, the sets $\{\rho: S(\rho) \leq n\}$ are closed, their complements are dense, hence they are nowhere dense and

$$\{\rho : S(\rho) < \infty\} = \bigcup_{n=1}^{\infty} \{\rho : S(\rho) \leq n\}$$

is of first category.
Besides lower semi-continuity, some other restricted continuity properties are valid. The most trivial one is

Convergence of canonical approximations[3]. If $\rho = \sum \rho^{(i)} P_i$, the $\rho^{(i)}$ being arranged in decreasing order, the P_i being one-dimensional, let

$$\rho_N = \sum_{i=1}^{N} \rho^{(i)} P_i \Big/ \sum_{i=1}^{N} \rho^{(i)}$$

("canonical approximation"). Then, $S(\rho_N) \to S(\rho)$.
Much less trivial is the

Dominated convergence theorem for entropy[9]. If ρ_n is a sequence of density matrices converging weakly to ρ, and if there exists a compact operator $A \geq 0$ (not necessarily a density matrix) such that $\rho_n \leq A$ for all n and $-\mathrm{Tr}\, A \ln A <$ $< \infty$, then $S(\rho_n) \to S(\rho)$.

Entropy Inequalities (One Space).

For all entropy inequalities, the reader is referred to the review article by Lieb[10].

Concavity. $S(\lambda\rho_1 + (1-\lambda)\rho_2) \geq \lambda\, S(\rho_1) + (1-\lambda)\, S(\rho_2)$ $(0 \leq \lambda \leq 1)$. (This also proves monotonicity with respect to mixing).
Proof: This is true indeed for every concave function $f \geq 0$. Let $\{\phi_i\}$ be an orthonormal basis of the Hilbert space that diagonalizes $\rho := \lambda\rho_1 + (1-\lambda)\rho_2$. Then,

$$\mathrm{Tr}\, f(\rho) = \sum f((\phi_i|\rho\phi_i)) \geq \lambda \sum f((\phi_i|\rho_1\phi_i)) + (1-\lambda) \sum f((\phi_i|\rho_2\phi_i)).$$

Now, $f((\phi_i|\rho_{1,2}\phi_i)) \geq (\phi_i|f(\rho_{1,2})\phi_i)$, hence $\sum f((\phi_i|\rho_{1,2}\phi_i)) \geq \mathrm{Tr}\, f(\rho_{1,2})$. For another proof, see Lieb[10]. Usually, concavity is considered to be one of the most important properties of entropy.

Concavity extends to the following inequality: let $\rho = \sum \lambda_i \rho_i$ ($\lambda_i \geq 0$, $\sum \lambda_i = 1$). Then,

$$\sum \lambda_i\, S(\rho_i) \leq S(\rho) \leq \sum \lambda_i\, S(\rho_i) - \sum \lambda_i \ln \lambda_i.$$

The term $-\sum \lambda_i \ln \lambda_i$ may be referred to as "mixing entropy".

It suffices to prove the r.h.s. only. Let us first assume that the ρ_i are one-dimensional projections P_i. Then,

$$\rho \prec \sum \lambda_i\, Q_i,$$

where the Q_i are also one-dimensional projections, but, in addition, are mutually orthogonal[4]. This is true since

$$\rho^{(1)} + \ldots + \rho^{(n)} = \sup \mathrm{Tr}\, \rho\, P,$$

where the sup is taken over all projections P of dimension $\leq n$. (Ky Fan's inequality[7]). Now this is $\geq \mathrm{Tr}\, \rho (Q_1 \vee \ldots \vee Q_n) \geq \lambda_1 + \ldots + \lambda_n$. For the general case, write

$$\rho_i = \sum_j \rho_i^{(j)}\, P_{ij},$$

the P_{ij} being the eigenprojections of ρ_i. Then,

$$\rho = \sum_{ij} \lambda_i\, \rho_i^{(j)}\, P_{ij},$$

hence $S(\rho) \leq -\sum_{ij} \lambda_i\, \rho_i^{(j)} \ln(\lambda_i\, \rho_i^{(j)}) = -\sum_{ij} \lambda_i\, \rho_i^{(j)} \ln \rho_i^{(j)} - \sum_i \lambda_i \ln \lambda_i =$
$= \sum_i \lambda_i\, S(\rho_i) - \sum_i \lambda_i \ln \lambda_i$.[4]

There is an equality on the r.h.s. if Ran ρ_i is orthogonal to Ran ρ_j for $i \neq j$.

Coarse-graining. The coarse-graining relations known from Uhlmann Theory give of course rise to the corresponding entropy inequalities.

Entropy Inequalities (Two Spaces)

Additivity. Let $H = H_1 \otimes H_2$. If $\rho = \rho_1 \otimes \rho_2$, then $S_\alpha(\rho) = S_\alpha(\rho_1) + S_\alpha(\rho_2)$ for all $\alpha \in [0,\infty]$, since the eigenvalues of ρ are $\rho_1^{(i)} \cdot \rho_2^{(k)}$, hence

$$S_\alpha(\rho) = (1-\alpha)^{-1} \ln \sum_{ik} (\rho_1^{(i)} \rho_2^{(k)})^\alpha = (1-\alpha)^{-1} \ln[\sum_i (\rho_1^{(i)})^\alpha \sum_k (\rho_2^{(k)})^\alpha] =$$

$$= S_\alpha(\rho_1) + S_\alpha(\rho_2).$$

Additivity expresses the fact that, if a system consists of two independent parts (which is mathematically expressed by the density matrix $\rho_1 \otimes \rho_2$), then the information about the whole system is just the sum of the informations about its parts.

Subadditivity. Now let ρ be a density matrix in H and let $\rho_1 := \mathrm{Tr}_{H_2} \rho$, $\rho_2 := \mathrm{Tr}_{H_1} \rho$ be the corresponding partial traces. By partial trace the following is meant: let $\{\phi_i\}$ be an orthonormal basis for H_1, $\{\psi_i\}$ be an orthonormal basis for H_2. Then the matrix elements of ρ_1, which is an operator in H_1, are given by

$$(\phi_i|\rho_1\phi_k) = \sum_j (\phi_i \otimes \psi_j|\rho(\phi_k \otimes \psi_j)).$$

This definition does not depend on the particular choice of $\{\psi_i\}$. It is easily checked that, for $A \in B(H_1)$,

$$\mathrm{Tr}\, \rho_1 A = \mathrm{Tr}\, \rho\, (A \otimes 1_{H_2}),$$

this property may also be used as a definition of the partial trace. Hence one can say that ρ_1 contains just all those informations of ρ that refer to the first subsystem only. The statement, that $S(\rho) \leq S(\rho_1) + S(\rho_2)$ is called "subadditivity".

Proof. Let $\{\psi_i\}$ be an orthonormal basis in H_2 that diagonalizes ρ_2. Then H may be written as $\bigoplus H_i$ with $H_i = H_1 \otimes \psi_i$. With respect to this decomposition, ρ has a matrix representation

$$\begin{pmatrix} \rho_{11} & \rho_{12} & \rho_{13} & \cdots \\ \rho_{21} & \rho_{22} & \cdots & \\ \cdots & & & \end{pmatrix}$$

ρ_1 is the density matrix $\sum \rho_{ii}$, whereas ρ_2 is the numerical matrix

$$\begin{pmatrix} \lambda_1 & 0 & \cdots \\ 0 & \lambda_2 & \\ \vdots & & \ddots \end{pmatrix}$$

with $\lambda_i = \mathrm{Tr}\, \rho_{ii}$. Now,

$$\rho \prec \begin{pmatrix} \rho_{11} & 0 & \cdots \\ 0 & \rho_{22} & \\ \vdots & & \ddots \end{pmatrix}$$

hence $S(\rho) \leq \sum \lambda_i\, S(\rho_{ii}/\lambda_i) - \sum \lambda_i \ln \lambda_i \leq S(\sum \rho_{ii}) - \sum \lambda_i \ln \lambda_i$ (concavity) = $= S(\rho_1) + S(\rho_2)$. For other proofs, perhaps more elegant, see 10, 11. Also $S_0(\rho)$ is subadditive, but no other α-entropy. Subadditivity certainly is one of the most important properties of entropy. It may be interpreted from the information-theoretical point of view in such a way that, if one takes the partial traces ρ_1 and ρ_2 and fits them together, all information about correlations is lost and, therefore, the entropy of $\rho_1 \otimes \rho_2$ (= $S(\rho_1) + S(\rho_2)$) must be bigger than the entropy of the original ρ.

<u>Monotonicity with respect to enlargening of the space</u>. This would be the statement that $S(\rho) \geq S(\rho_1)$. Although this is true in the classical case, it is false in the quantum case since ρ may be pure, but ρ_1 may not. In fact, to every density matrix ρ_1 one can find a Hilbert space H_2 and a pure density matrix ρ in $H_1 \otimes H_2$ with $\rho_1 = \mathrm{Tr}_{H_2} \rho$. By the way, in that case, the positive spectra of ρ_1 and ρ_2 coincide, hence $S(\rho_1) = S(\rho_2)$[12].

<u>Triangle inequality</u>[12]. $|S(\rho_1) - S(\rho_2)| \leq S(\rho) \leq S(\rho_1) + S(\rho_2)$.

The r.h.s. being subadditivity, one has to prove the l.h.s. only. Let H_3 be a Hilbert space and ρ' be a pure density matrix in $H_1 \otimes H_2 \otimes H_3$ such that $\rho = \mathrm{Tr}_{H_3}\rho'$. Let $\rho_3 = \mathrm{Tr}_{H_1 \otimes H_2}\rho'$. Then, $S(\rho) = S(\rho_3)$. $S(\rho_1) = S(\rho_{23})$, where $\rho_{23} := \mathrm{Tr}_{H_1}\rho'$. Subadditivity yields

$$S(\rho_1) = S(\rho_{23}) \leq S(\rho_2) + S(\rho_3) = S(\rho_2) + S(\rho)$$

and interchanging of 1 and 2 proves the triangle inequality.

<u>Entropy Inequalities (Three Spaces)</u>.

This is a group of inequalities centered around strong subadditivity. Strong subadditivity means the following: let $H = H_1 \otimes H_2 \otimes H_3$, ρ be a density matrix in H, $\rho_{12} = \mathrm{Tr}_{H_3}\rho$, $\rho_{23} = \mathrm{Tr}_{H_1}\rho$, $\rho_2 = \mathrm{Tr}_{H_1 \otimes H_3}\rho$. Then,

$$S(\rho) + S(\rho_2) \leq S(\rho_{12}) + S(\rho_{23}).$$

If H_2 is one-dimensional, this reduces to subadditivity.
This is a highly non-trivial result and a proof of it requires inequalities that are very hard to derive, for instance that (for finite-dimensional matrices) the mapping

$$A \to \mathrm{Tr}\, e^{K + \ln A}$$

is concave[14]. Since this is a field full of technicalities, we have to refer the reader to the literature 10, 13, however, it should be pointed out that strong additivity has various important implications in physics.

<u>Axiomatic Characterizations</u>[3]

Let Φ be a mapping of the set of density matrices into $[0,\infty]$. One may ask which conditions have to be imposed on Φ in order that, up to a constant factor, Φ is the entropy.

Preliminary axioms. (P1) $\Phi(\rho)$ is finite, if ρ is of finite rank.
(P2) If ρ is not of finite rank, then $\Phi(\rho_N) \to \Phi(\rho)$, where the ρ_N are the canonical approximations of ρ.
(P3) Φ fulfils partial isometric invariance.

Characterization "à la Renyi". Let Φ fulfil (P1) - (P3) and
(R) If $H = H_1 \oplus \ldots \oplus H_n$, and $\rho = \lambda_1\rho_1 \oplus \ldots \oplus \lambda_n\rho_n$ (ρ_i being density matrices, $\lambda_i \geq 0$, $\sum \lambda_i = 1$), then $\Phi(\rho) = \sum \lambda_i \Phi(\rho_i) + \Phi(\Lambda)$, where Λ is a density matrix in the Hilbert space C^n with eigenvalues $\lambda_1,\ldots,\lambda_n$. (The H_i need not have the same dimension).
Then, $\Phi(\rho) = \text{const}\cdot S(\rho)$.
For the proof, it suffices to consider the case that all ρ_i are of finite rank. Choosing suitable orthonormal bases in the H_i, and assuming that all density matrices under consideration commute, one is left with the classical situation[2] (since (P3) implies symmetry and expansibility), hence $\Phi(\rho) = \text{const}\cdot S(\rho)$, a fortiori this is true for all ρ.

Characterization "à la Aczel, Forte, and Ng". Let Φ fulfil (P1) - (P3) and additivity as well as subadditivity. Then Φ is a linear combination of $S(\rho)$ and $S_o(\rho)$.
Proof. Again it is fairly simple to reduce the situation to the classical case. Then one can apply the very remarkable theorem of Aczel, Forte, and Ng[15]: let ψ be a function, defined for all finite probability distributions (i.e. for all n-tuples $(p_1,p_2,\ldots,p_n)$ such that $p_i \geq 0$, $\sum p_i = 1$) with the properties

(i) $\psi \geq 0$

(ii) $\psi(p_{P(1)},\ldots,p_{P(n)}) = \psi(p_1,\ldots,p_n)$, where P is any permutation of (1,..,n) ("symmetry")

(iii) $\psi(p_1,\ldots,p_n,0) = \psi(p_1,\ldots,p_n)$ ("expansibility")

(iv) $\psi(p_1q_1,\ldots,p_nq_m) = \psi(p_1,\ldots,p_n) + \psi(q_1,\ldots,q_m)$ ("additivity")

(v) $\psi(r_{11},\ldots,r_{nm}) \leq \psi(p_1,\ldots,p_n) + \psi(q_1,\ldots,q_m)$, where
$p_i = \sum_j r_{ij}$, $q_i = \sum_j r_{ji}$ ("subadditivity").

Then, $\psi(p_1,\ldots,p_n) = - A \sum p_i \ln p_i + B\cdot$(logarithm of the number of p's $\neq 0$), with A and B independent of n.

The proof of this theorem in principle is elementary except for one number-theoretical argument, but very tricky.

It remains to eliminate the quantum Hartley entropy $S_0(\rho)$. This can be done by several very mild continuity conditions, for instance of that kind: let ρ_n be a sequence of density matrices of rank 2, P be a one-dimensional projection, $[\rho_n, P] = 0$, $||\rho_n - P|| \to 0$, then $S(\rho_n) \to 0$ (or one could even demand only that $S(\rho_n) \to S(P)$, which is not clear a priori)[3].

Of course, the characterization "à la Aczel, Forte, and Ng" is much more related to physics than the one "à la Renyi" since additivity and subadditivity have a rather appealing physical interpretation.

Acknowledgments

The author wishes to thank Prof. Elliott H. Lieb for a critical reading of the manuscript and for making numerous suggestions, as well as Prof. O.E. Lanford III for useful remarks.

References

1. J.v.Neumann, Z. Phys. 57, 30 (1929)
2. A. Renyi, Wahrscheinlichkeitsrechnung, Deutscher Verlag der Wissenschaften, Berlin 1966
3. W. Ochs, Rep. Math. Phys. 8, 109 (1975)
4. A. Uhlmann, Wiss. Z. Karl-Marx-Univ. Leipzig, Math.-Naturwiss. R. 20, 633 (1971); 21, 421 (1972); 22, 139 (1973); private communications
5. A. Wehrl, Rep. Math. Phys. 6, 15 (1974)
6. G. Hardy, J. Littlewood, G. Polya, Inequalities, Cambridge, 1967
7. E. Beckenbach, R. Bellmann, Inequalities, Springer 1971
8. A. Wehrl, Rep. Math. Phys., to be published
9. B. Simon, appendix to ref. 12
10. E. Lieb, Bull. AMS 81, 1 (1975)
11. D. Ruelle, Statistical Mechanics, Benjamin, 1969
12. H. Araki, E. Lieb, Comm. Math. Phys. 18, 160 (1970)
13. E. Lieb, M. Ruskai, J. Math. Phys. 14, 1938 (1973)
14. E. Lieb, Adv. Math. 11, 267 (1973)
15. J. Aczel, B. Forte, C. Ng, Adv. Appl. Prob. 6, 131 (1974)